SANDSTONE LANDFORMS

Sandstones form the backdrop to some of the world's most spectacular scenery – forming high mountain ranges, bold cliffs, extensive plateaus, impressive caverns and magnificent towers. They are found all over the planet and in all climates, from hot deserts to the polar region, and provide the construction material for iconic buildings in numerous countries.

Following on from the authors' successful 1992 book, this is the only volume that considers sandstone landforms from a truly global perspective. It describes the wide variety of landforms that are found in sandstone, and discusses the role of lithological variation, chemical weathering and erosional processes in creating these features, with examples drawn from around the world. Climatic and tectonic constraints on the development of sandstone landscapes are also considered.

This volume provides a comprehensive assessment of the literature from publications in a range of languages, and is illustrated with over 130 photographs of sandstone features from every continent. It presents a holistic account of sandstone terrain for researchers and graduate students in a variety of fields including geography, geomorphology, sedimentology and geomechanics.

ROBERT W. YOUNG received a Ph.D. from the University of Sydney in 1975, and was based for over 20 years in the Earth and Environmental Sciences department at the University of Wollongong, Australia.

ROBERT A. L. WRAY received a Ph.D. from the University of Wollongong in 1996, and is an Honorary Fellow in the Earth and Environmental Sciences department at the same university.

ANN R. M. YOUNG received a Ph.D. from the University of Wollongong in 1983, and was based for 20 years in the Earth and Environmental Sciences department at the same university.

Collectively, the authors have over 70 years' experience in the subject.

SANDSTONE LANDFORMS

ROBERT. W. YOUNG
formerly University of Wollongong, Australia

ROBERT. A. L. WRAY
University of Wollongong, Australia

ANN. R. M. YOUNG
formerly University of Wollongong, Australia

CAMBRIDGE
UNIVERSITY PRESS

CAMBRIDGE
UNIVERSITY PRESS

University Printing House, Cambridge CB2 8BS, United Kingdom

One Liberty Plaza, 20th Floor, New York, NY 10006, USA

477 Williamstown Road, Port Melbourne, VIC 3207, Australia

314-321, 3rd Floor, Plot 3, Splendor Forum, Jasola District Centre, New Delhi - 110025, India

79 Anson Road, #06-04/06, Singapore 079906

Cambridge University Press is part of the University of Cambridge.

It furthers the University's mission by disseminating knowledge in the pursuit of education, learning and research at the highest international levels of excellence.

www.cambridge.org
Information on this title: www.cambridge.org/9781108462044

© R. Young, R. Wray and A. Young 2009

First published 2009
First paperback edition 2018

A catalogue record for this publication is available from the British Library

ISBN 978-0-521-87733-6 Hardback
ISBN 978-1-108-46204-4 Paperback

Contents

Contents

Preface

Sandstones are found worldwide – from Greenland to Antarctica, and on all continents. The Old Red Sandstone forms the spine of Britain, extending from Wales to the Orkney Islands. Its stratigraphy and fossils were studied by the doyen of geology, James Hutton, and were the core of the 'map that changed the world' by the pioneering geologist, William Smith. The world's tallest waterfall, Angel Falls, tumbles over the sandstones of the Roraima in Venezuela. Iconic buildings in many parts of the world are made of sandstone. Movie-goers will recognize the sandstone terrain of Utah in many Westerns and the sandstone spires in the Czech Republic in the film *The Lion, The Witch and The Wardrobe*. Sandstone and its landforms are therefore of interest not only academically but generally.

The Youngs' earlier book, *Sandstone Landforms*, was published by Springer in 1992 (copyright reverted to Robert and Ann Young in 1997). Their aim then was to draw together the main explanations of sandstone geomorphology from accounts scattered throughout the literature, and to add their own field observations. This was written as a high-level academic book, and is now out of print. To our knowledge, its only predecessor was a 1972 treatise in French by Monique Mainguet. Since 1992, the focus of research on sandstones has shifted to what was then a new field – the widespread and significant impacts of silica solution on sandstone landscapes. A major review of European sandstones is due for publication (Hartel *et al.*, in press). Still however, much information remains scattered within a plethora of scientific journals.

Our aim here is to bring together not just an overview of research on sandstone landforms, but a global perspective that includes assessment of the underlying principles used in interpreting the landscapes. We have updated and revised the previous work, and the section on solutional landforms and processes has been greatly expanded. As with the earlier edition, we hope that we can convey some of the fascination and interest – some may even say, absorbtion – that sandstone landscapes have provided for all of us.

We are grateful to colleagues who have provided photographs (as acknowledged in the captions) and assistance – Piotr Migon, Stefan Doerr, Rowl Twidale, Dennis Netoff, Andy Spate and John Dixon. We acknowledge also the use of photographs from Mr Hong Kaidi, Jan Galloway and the estate of the late J. N. Jennings. At Cambridge University Press, London, we thank Matt Lloyd for arranging the publication, and Diya Gupta, Anna-Marie Lovett and Sarah Lewis for editorial assistance.

1

Introduction

Given the great emphasis placed on sands and sandstones in sedimentological research, the relative neglect of the geomorphology of sandstones seems paradoxical. It seems all the more so because much of the pioneer work in modern geomorphology, from James Hutton's work in 1795 onwards, came from sandstone landscapes. Hutton's recognition of the interplay of tectonic forces and of prolonged and repeated fluvial erosion in the shaping of the surface features of the Earth was based on observations at places like Siccar Point in Scotland, where it is obvious that Silurian greywackes were folded and extensively eroded before they were buried beneath the Devonian Old Red Sandstone. And Hugh Miller's *The Old Red Sandstone*, which appeared in seven editions between 1841 and 1889, contains perceptive and evocative accounts of how changes in the properties of the sandstone give rise to distinctive types of landforms:

We pass from the conglomerate to the middle and upper beds of the lower formation, and find scenery of a different character in the districts in which they prevail. The aspect is less bold and rugged, and often affects long horizontal lines, that stretch away, without rise or depression, amid the surrounding inequalities of the landscape, for miles and leagues, and that decline to either side, like roofs of what the architect would term a low pitch.

(Miller, 1889, p. 214)

It was also in sandstone lands that J. D. Dana refuted Charles Darwin's marine hypothesis for the origin of the great valleys in the Blue Mountains of New South Wales. Darwin, publishing in 1839 and then in subsequent editions, argued that 'To attribute these hollows to the present alluvial action would be preposterous' (Darwin, 1839, in the 1890 edn, p. 319). Drawing on his own observations during the voyage of the *Beagle*, and also on Lyell's

1

advocacy of marine erosion, he proposed that these broad valleys, which leave the upland through narrow gorges, are primarily the products of marine deposition:

I imagine that the strata were heaped by the action of strong currents, and of undulations of an open sea, on an irregular bottom; and that the valley-like spaces thus left have their steeply sloping flanks worn into cliffs, during the elevation of the land; the worn-down sandstone being removed either at the time when the narrow gorges were cut by the retreating sea, or subsequently by alluvial action.

(Darwin, 1839, in the 1890 edn, p. 320)

Although Dana had come to a similar conclusion during his visit to New South Wales in 1840, he later concluded that Thomas Mitchell, the colonial surveyor and explorer, had been correct in attributing these valleys to erosion by streams:

The idea that running water was the agent in these operations appears not so 'preposterous' to us, as it is deemed by Mr. Darwin; and we think that it may be shown that Major Mitchell was right in attributing the effect to this cause. The extent of the results is certainly no difficulty with one who admits time to be an element which a geologist has indefinitely at his command.

(Dana, 1850, p. 289)

Ironically, in the same year as Dana's paper was published, J. B. Jukes, who later became the great British advocate of fluvial erosion, quoted Darwin's descriptions in likening the valleys of these same sandstone lands to winding harbours (Jukes, 1850, pp. 24–25).

That erosion by streams is the dominant process in shaping surface features, even in arid and semi-arid lands, was demonstrated during the latter part of the nineteenth century by outstanding research on the Colorado Plateau of North America. Much of this pioneering work (e.g. Dutton, 1882; Powell, 1895) was carried out in the extensive sandstone lands of the plateau. In Dutton's words:

It would be difficult to find anywhere else in the world a spot yielding so much subject matter for the contemplation of the geologist; certainly there is none situated in the midst of such dramatic and inspiring surroundings.

(Dutton, 1882, p. 92)

Nonetheless, important work on the impact of climatic change and on the role of seepage and solution was being carried out in the less dramatic sandstone landscapes of Europe. The work of Hettner (1887, 1903) in the 'Saxon Switzerland' of the Elbe River in Germany is a case in point, especially with regard to seepage and basal sapping of sandstone outcrops. And it should be noted that the present-day focus on solutional processes in sandstone was foreshadowed by the French geographer de Martonne who, in his a *Shorter*

Physical Geography (1927, p. 171), included the solution in sandstone under the heading of 'Origin of Karst Topography'.

Research methodology

Although important papers on sandstone landforms have appeared from time to time (e.g. Gregory, 1917, 1950; Schumm and Chorley, 1964, 1966; Mainguet, 1972), research in this field seems almost insignificant when compared with the vast output from the study of landforms on granites and, even more so, on limestones. And although interest in sandstone landforms has increased during the last decade or so, attention has been focused very much on the issue of solutional features. The disparity is not due to the sandstones themselves being spatially insignificant, for they occupy approximately the same proportion of the continents as do granites and carbonates (Meybeck, 1987). Nor is it because sandstones do not form interesting or spectacular landscapes; in a recent major book on the great geomorphological landscapes of the world, nearly a quarter are formed in sandstones (Migon, in preparation). Rather, it probably lies in the widespread assumption that the natural sculpturing of gently dipping and resistant sandstones is a relatively straightforward matter that holds few surprises. This assumption is strikingly illustrated in the selection of sandstone by Tricart and Cailleux (1972) as the ideally simple medium for demonstrating the dominant role of climate in morphological diversity. Yet lithological and structural simplicity must be demonstrated rather than assumed, for our understanding of the interaction of surface processes and the properties of rocks is far from complete. Indeed, the diversity of sandstone landscapes is the major theme of this book.

The pitfalls of premature generalization from relatively few case studies are highlighted further by the diversity of the engineering, or rock mechanics properties of sandstones. Intact rock is generally rated by Deere–Miller plots of the uniaxial compressive strength and the modulus of elasticity (Lama and Vukuturi, 1978). Some sandstones fall entirely into the very low strength group (Dobereiner and de Freitas, 1986), while others, such as the highly quartzose Proterozoic sandstones of the Kimberley region of northern Australia (R. W. Young, 1987), fall at the other end of the range in the very high strength group. Plots for just a single formation, the Hawkesbury Sandstone of the Sydney Basin in southeastern Australia, give a range from very low strength to high strength (Pells, 1977, 1985). Even the mechanisms of fracturing change dramatically from the shattering of individual grains in highly indurated sandstones to the rolling or displacement of grains in weak, poorly cemented sandstones (Dobereiner and de Freitas, 1986). Indeed, in the case of the so-called 'flexible sandstones', or 'itacolumites' (Dusseault, 1980), the lack of true cement, coupled with a tight interlocking of grains, allows the rock to bend and recover under conditions in which brittle failure would normally be expected.

The definition of sandstone adopted in this book includes orthoquartzites, which are sandstones so completely and strongly cemented by secondary quartz that they break across the grains rather than through the cement. Although it is desirable on petrographic grounds to distinguish orthoquartzites from meta-quartzites produced by metamorphic recrystallization of quartz, it is not always practical to do so in geomorphological studies because some orthoquartzites and metaquartzites have similar mechanical properties. Furthermore, it is often difficult to determine whether the term 'quartzite' used in some geomorphological studies refers to orthoquartzites or metaquartzites. Where possible we draw attention to the way in which such terms have been used in particular studies. And, as is illustrated specifically by examples later in this chapter, we emphasize the geomorphogical differences between orthoquartzites and less strongly cemented sandstones.

Because of the relative neglect of the geomorphology of sandstones there is no soundly established methodology, like that of research on karst, which can be followed. Attempts to structure research in this field on the basis of the supposed dominance of climatic controls (Ahnert, 1960; Tricart and Cailleux, 1972) seem premature, especially in the light of the compelling evidence of major structural and lithological influences (Bradley, 1963; Robinson, 1970; Oberlander, 1977). This is not to deny a role to climate – a role which has been strikingly demonstrated by Mainguet (1972) – but simply to refuse it pride of place, and not to relegate lithology and structure to a secondary or 'passive' role (cf. Büdel, 1982). Notwithstanding excellent work such as Mainguet's investigation of chemical weathering of sandstones, much of the research into climatic influences has been 'merely a systematization of simple observations' (Yatsu, 1966, p. 3) that has little true explanatory power. How then should we attempt to explain sandstone landforms?

Scientific explanation, as Bateson (1973, pp. 26–27) so succinctly expressed it, 'is the mapping of data on to fundamentals'. To this definition he added that 'in scientific research you start from two beginnings, each of which has its own kind of authority: the observations cannot be denied, and the fundamentals must be fitted. You must achieve a sort of pincers manoeuvre'. Moreover, Bateson warned against mistaking loosely defined explanatory notions, or heuristic devices, for the precisely conceptualized building blocks of science. A similar, if more vitriolic, critique has been directed by Yatsu (1966, 1988) specifically against the reliance on such heuristic devices in much geomorphological research, which he likens to *Ikebana* (flower arranging). Yatsu's great contribution, however, has been to delineate, in detail, the fundamental concepts on which our explanations of landforms should be based. Here we have followed his lead, and have turned to fundamental concepts such as strength, stress, strain and reaction rates, developed in the allied fields of rock mechanics, silicate weathering and fluid dynamics. Nevertheless, it is a 'pincers manoeuvre' that is required, not just an application of basic concepts. The study of landforms is more than just applied geophysics and geochemistry, for selecting those phenomena which are significant to geo-morphology requires an appreciation not only of theory, but more so of actual landscapes. In outlining the complexity of the task of explaining sandstone landscapes, we therefore begin with field observations (Fig. 1.1) rather than with a survey of fundamentals.

A study of variations in sandstone landforms – the East Kimberley region, Western Australia

The East Kimberley region of northwestern Australia (location 12, Fig. 1.1) con-tains a vast area of generally gently dipping Proterozoic and Paleozoic sandstones. These sandstones have been continuously denuded, apparently under seasonally humid tropical climates, since early in the Tertiary or late in the Mesozoic. Despite their uniform tectonic and denudational setting, the present-day topography of these sandstones is remarkably varied, even over distances of only a few kilometres (R. W. Young, 1987, 1988). Many of them have been carved into cliff-lined mesas; others have been sculptured into great domes; and some have been intricately dissected into labyrinthine complexes of towers, pinnacles and sinuous, knife-edged ridges (Figs 1.2–1.5). Explaining this diversity is no simple task.

The problem of diversity within a region of essentially uniform present-day climate is not resolved by appeals to past climatic change, especially to the expansion or contraction of the desert which lies to the south of the Kimberleys. The angular clifflines of the Cockburn Range, that seem similar to the standard models for arid regimes, are on the humid side of the region; whereas the domes and convex towers of the Bungle Bungle Range, which are more like the rounded forms supposedly typical of humid regimes, are on the arid margins. Nor is there any evidence to suggest that the angular and rounded forms represent differing generations of landforms. On the contrary, the field evidence points to continuity in the style of sculpture. Where a distinct break separating two generations of landforms can be recognized between the Cainozoic summit erosional surface and the modern flanks of the Bungle Bungle Range, essentially the same array of towers and ridge development is found on each (R. W. Young, 1987). What is more, the lateral transition from towers to domes in the Bungle Bungle Range that occurs over a few kilometres along continuous escarpments can in no way be attributed to a climatic origin, past or present. Thus, in recognizing the degree of

Fig. 1.1 Location of major sandstone landscapes considered in this book.
1 Ellsworth Mts; 2 Prince Charles Mts; 3 McMurdo Dry Valleys; 4 Torlesse Range; 5 Tararua Range; 6 Sydney Basin; 7 Grampian Range; 8 Flinders Range; 9 Uluru; 10 Stirling Range; 11 Murchison Gorge; 12 East Kimberley; 13 Arnhemland; 14 Carnavon Range; 15 Torricelli Mts; 16 Taiwan; 17 Danxiashan; 18 Wanfoshan; 19 Chishui; 20 Brooks Range; 21 central Canadian Rocky Mts; 22 Gros Ventre; 23 Zion Canyon; 24 Valley of Fire; 25 Monument Valley; 26 Wisconsin Driftless Area; 27 northern Appalachian Mts; 28 Roraima; 29 Chaco Basin; 30 Minas Gerais; 31 Vila Velha; 32 Nuussuaq Basin; 33 Keyser Franz Joseph Fjord; 34 Svalbard; 35 Varanger Peninsula; 36 Rondane Mts; 37 Hornelen & Solund Basins; 38 Orkney; 39 Torridonian Mts; 40 Pennines; 41 Fontainebleau; 42 Catalan Ranges; 43 Saxony-Bohemia; 44 Stolwe Mts; 45 Meteora; 46 Bashkiria; 47 Jordan; 48 Msak Mallat; 49 Tibesti; 50 Borkou; 51 Moroccan High Atlas Mts; 52 Adrar; 53 Fouta Djallon; 54 SE Nigeria; 55 Table Mt; 56 Clarens Valley; 57 East Transvaal; 58 Chimanimani Highlands; 59 Isalo; 60 Kialas; 61 Bhutan Himalayas

Fig. 1.2 Cliffs in the Cockburn Range, Western Australia

Fig. 1.3 Joint-bounded domes in conglomerates and pebbly sandstones in the Bungle Bungle Range, Western Australia

Fig. 1.4 Rounded towers cut in friable sandstones in the Bungle Bungle Range, showing marked banding on the rock surfaces

Fig. 1.5 Sinuous narrow ridges cut in friable sandstones rise abruptly from the surrounding plain, Bungle Bungle Range

diversity among these landforms, we encountered – at the very outset of our considerations – a methodological stumbling block to any attempt to fit this region into the morphogenetic systems proposed for sandstone geomorphology. Those systems are based on the comparative analysis of apparently representative type examples. But which is the type example here – the cliffs, the domes, or the towers and narrow ridges? This question cannot be dismissed by simply appealing to the scale of observation at which climatic or structural constraints are supposedly most discernible; that is to say, we are not just dealing with structurally controlled deviations from a common morphogenetic type. We have encountered here a major problem in the very rationale of the climatic classification of landforms itself, for the variety of forms within this region is greater than that found between the supposedly major morphogenetic zones (cf. Tricart and Cailleux, 1972). Whether the Kimberleys are simply the exception to a valid general rule, or whether the diversity of sandstone landforms within other morphogenetic zones really does rival the diversity between zones, is a vital issue to which we will return.

Likewise, the dominance of any other single factor, even if demonstrated elsewhere, cannot be assumed to apply here. Structure is a case in point. The constraints on topography imposed by variable jointing and bedding are expressed clearly on many outcrops in the region. Nonetheless, the domes of the Kimberley region do not display the beautifully curved unloading or spalling planes which characterize many of their counterparts on the Colorado Plateau (Bradley, 1963); nor is the preferential excavation of joints normally associated with stream erosion of sandstones displayed to any appreciable extent in the dissection of most tower and ridge complexes. On the contrary, some of the joints cutting through these complexes are case hardened, and act as lines of resistance to erosion (R. W. Young, 1986).

The role of lithology

Lithology appears to be a far more important constraint on landforms here than does structure. The cliff-lined mesas are best developed in the Proterozoic sandstones; the domes occur in Paleozoic pebbly sandstones and conglomerates; and the tower and ridge complexes are limited almost entirely to well-bedded Paleozoic sandstones. Yet, as Yatsu (1966) has emphasized, simple correlation of topography with lithological types explains little, for it is the variable mechanical properties of the rocks that must be assessed.

Most of the Proterozic sandstones in the region are strongly cemented, and some are very tough indeed. The Cockburn Sandstone, for example, has a compressive strength of around $100\,\text{MPa}$ ($1\,\text{megaPascal} = 145\,\text{lb/in}^2$). As this massive siliceous sandstone – most outcrops of which can be classified as

Introduction

Fig. 1.6 Thin section of Cockburn Sandstone, showing high grain-to-grain contact

orthoquartzite – also has horizontal bedding and widely spaced, vertical jointing, it can be expected to stand in long-term equilibrium slopes of about 70° (cf. Selby, 1982) and, especially when undercut, to form high vertical faces. In contrast to the very tough Cockburn Sandstone, most of the Paleozoic sandstones of the region are generally more friable and easily eroded, yet still stand in quite steep, occasionally vertical, slopes. These topographic characteristics reflect the curious mechanical properties of these Paleozoic sandstones. Even the most friable of them, like the Glass Hill Sandstone that forms the Bungle Bungle Range, generally have reasonably high strengths of 40–60 MPa when in compression, but have such low strengths when tensional or shearing stresses are applied that they can be broken by hand pressure.

The reasons for the variable mechanical properties of the Kimberley sandstones can be readily seen at the microscopic level. The very tough ones, like the Cockburn Sandstone, have a high percentage of grain-to-grain contact, and are highly indurated with quartzose cement (Fig. 1.6). Friable ones, like the Glass Hill Sandstone (R. W. Young, 1988), are also composed mainly of angular, closely interlocking grains with a high proportion of grain-to-grain contact, but have virtually no cement (Fig. 1.7). In this case the interlocking fabric of grains can carry considerable compressive stresses, but individual grains can be detached by very low shearing stresses. Hence, steep slopes can still be maintained in material which is easily eroded.

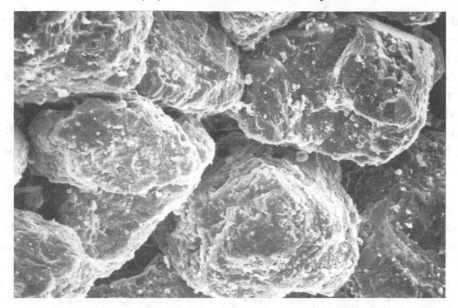

Fig. 1.7 SEM image of the Glass Hill Sandstone, showing etching of grains and lack of cement

Microscopic study not only explains the contrasting present-day mechanical properties of the Kimberley sandstones; it also shows how these contrasts, which were once much less striking than they are now, have developed. The now friable sandstones were originally quite well indurated, for they contain numerous remnants of quartz overgrowths that once formed sutures between the grains. These overgrowths were subsequently extensively etched and the bonds between most grains destroyed. Two major questions concerning the chemical alteration of the sandstones now loom large. Why were grain sutures destroyed in the Paleozoic sandstones to a far greater degree than in their Proterozoic counterparts? And under what environmental conditions did such large-scale etching of quartz occur?

The first question is the more readily answered, but requires consideration of the contrasting diagenetic histories of the sandstones, a topic rarely dealt with in geomorphological studies. The induration of most of the once deeply buried Proterozoic sandstones was so great that primary porosity was almost completely eliminated, thereby precluding the later movement of corrosive solution through the rock as a whole (see Fig. 1.6). Very substantial etching did occur where such solutions flowed along major fractures, for these highly quartzose rocks now contain extensive caves developed along intersecting joint planes (Jennings, 1983; R. W. Young, 1987). Nonetheless, the rock between the major fractures remained virtually unaltered. In the Paleozoic sandstones, however, microscopic studies (R. W. Young, 1988) showed that primary porosity was not eliminated;

and that corrosive solutions were thus not limited to major fractures, but were able to diffuse through the rock, destroying intergranular cement and sutured quartz overgrowths (see Fig. 1.7). In the case of the Glass Hill Sandstone, the destruction was almost complete, leaving a mass of uncemented, interlocking quartz grains with clay in some interstices.

Microscopic analysis also indicates the origin of the very distinctive orange and grey banding on outcrops of the Glass Hill Sandstone. These bands parallel the bedding, and can be traced from outcrop to outcrop for kilometres. Individual bands are up to several metres wide, and although from a distance their boundaries seem sharp, on closer inspection the contacts between them are irregular. Of particular significance is that the banding is only a surface phenomenon, for the sandstone is uniformly bleached beneath grey and orange bands alike. The banding seems to be due to slight variations in the porosity and clay content between beds (R. W. Young, 1986). Traces of iron impart the orange hues in a manner similar to weakly developed desert varnish. Cyanobacteria impart the grey hues to beds that remain moist for longer periods because of higher clay content (Hoatson *et al.*, 1997). Both types of banding constitute a surface case-hardening that retards erosion of the friable grains beneath.

The role of weathering

Delineating the conditions which gave rise to the extensive leaching of silica, like that of the Glass Hill Sandstone in the Bungle Bungles, on a regional scale is a much more difficult issue to resolve. As is generally argued (Pouyllau and Seurin, 1985; Meybeck, 1987), a very great duration of seasonally humid tropical climates seems the most obvious cause. It must be remembered, however, that the link between field evidence and this hypothesis of the leaching of silica under highly acidic conditions is largely inferential, and has yet to be reconciled with laboratory studies which demonstrate greater solubility of silica under conditions of high, rather than low, pH. Furthermore, the presence of microscopic salt crystals on some highly etched grains of the Glass Hill Sandstone (R. W. Young, 1988) suggests that factors other than the temperature and pH of the solution may be crucial in the leaching of silica. This suggestion is supported by laboratory work, and by field evidence from other parts of Australia (A. R. M. Young, 1987; Wray, 1997c). Whatever the precise cause, observations such as those in the Kimberleys have demonstrated beyond any doubt that large-scale and widespread chemical corrosion can occur in highly quartzose rocks. In short, we are now faced with the paradox of karst landforms, normally associated with the chemically weakest of rocks, being extensively developed in some of the most chemically resistant of rocks (Wray, 1997a).

Differential weathering explains much of the varying degree and style of erosion of sandstones in the Kimberley region. Whereas drainage lines cut in the Cockburn Sandstone are widely spaced and show strong structural constraint, those in the friable Glass Hill Sandstone are more closely spaced and have a general dendritic pattern that displays much less structural constraint. However, in the central part of the Bungle Bungle Range, the dendritic drainage and towered landforms typical of the Glass Hill Sandstone give way to rectilinear drainage and towering cliffs. This transition is probably the result of subtle metamorphism associated with the Piccaniny Circular Structure which is considered to be probably the remnant of a meteor impact.

The extremely abrupt break of slope between the steep valley walls and the flat valley floors or flanking pediments of the Bungle Bungle Range (see Figs 1.4, 1.5) seems at first sight to mark the contact between friable and relatively unweathered Glass Hill Sandstone, especially as equally impressive breaks of slope have undoubtedly been formed by the stripping of weathering fronts in granite a little further to the west in the Kimberleys (Whitaker, 1978). Again, however, caution is needed in applying general principles. In fact, the friable sandstone extends well below the level of abrupt change in gradient around the Bungle Bungle Range, and the extensive pediments have formed not by the stripping of a weathering front, but by the extension of an erosional plane from the main drainage lines into the weathered sandstone. This is true also of the ancient pediment preserved on the summit of the range. It lies on rocks that are no less weathered than those in an upper storey of towers into which it encroached, or those in the younger towers which developed in the rocks below, after the summit pediment was dissected. The concept of the stripping of weathered material from surfaces etched into hard rock is of no help here, and the origins of these erosional surfaces must be sought rather in the denudational and tectonic history of the region.

The distribution of the contrasting types of landforms that we have considered here reflects to a large extent the characteristics of the sedimentary basins in which the sandstones were deposited. It also reflects the subsequent diagenetic transformation and tectonic deformation of the sandstones. The effect of changes in depositional environments is clearly illustrated in the transition from domes to towers and ridges in the Bungle Bungle Range, for the conglomerates and pebbly sandstones in the west were closer to the erosional source of the sediments than were the finer sandstones further east. As we have already seen, the marked differences between the diagenetic histories of the Proterozoic and Paleozoic sandstones largely determined the differential response of these rocks to subsequent weathering, which in turn largely determined their mechanical properties. Despite their great age, the Cockburn and Glass Hill Sandstones have undergone little deformation; they are still essentially flat-lying. Separating them, however, is a zone up to 50 km wide in which there has been great deformation. Sandstones in that zone are steeply tilted; and because transcurrent and normal faulting has differentially displaced them, they have a discontinuous outcrop and give rise to rugged terrain.

Conclusions

Although an understanding of the distribution of sandstones and of the variations in their strength and resistance to weathering and erosion is essential, it provides only a partial answer to the question of how their surface forms developed. The interaction of the surface forms with the variable stresses operating on them must also be assessed, but it is precisely on this issue that most studies of landforms are weakest. The great majority of geomorphological textbooks do not even mention the geometry of stress patterns, and blithely ignore the well-established principles in the neighbouring fields of rock mechanics, structural engineering and architecture. Yet, as can readily be seen from vistas of the East Kimberley sandstones, landforms can be resolved into regularly recurring and clearly recognizable 'architectural' elements or structures, such as rock walls, domes, arches, buttresses and beams. How natural structures such as these sustain the loads imposed on them needs to be understood. Indeed, the varied morphologies of sandstones such as these offer probably the most promising setting for developing the study of the statics of landforms, a study to which the work of Gerber and Scheidegger (1973) and especially Yatsu (1988) points the way.

The purpose of this brief survey of landforms of the East Kimberley region has been to illustrate the character and magnitude of the task of developing a systematic geomorphology of sandstones. The first step must be to recognize the diversity of sandstone topography, rather than to arrange supposedly representative type examples into a genetic classification that is derived not from specifically defined problems, but from the *a priori* assumption of the dominance of a single factor. Such categorizing is likely to result in the fallacy of misplaced concreteness, whereby the categories are seen as fundamental entities, rather than as labels for a particular point of view. If the examples listed here show anything, it is that investigation must proceed in a pincers manouevre, in which the reality of field observations, no matter how inconvenient, is linked to fundamentals.

In recent years, considerable effort has been spent in categorizing sandstone landforms. A classification of sandstone (*Danxia*) landforms, consisting of 38 subdivisions over a wide range of sizes, has been devised by Chinese geomorphologists (Peng, 2000). And Mikuláš (2008a) has proposed a genetic

Table 1.1 *Varying scales of landforms developed on sandstones*

Scale	Selected forms
1	Etch pits or overgrowths on grains
2	Tafoni, weathering pans, tesselated surfaces, tors, runnels
3	Cliffs, benches, domes, towers, arches, dells, slot canyons, waterfalls, caves, dolines
4	Ruiniform assemblages, karst assemblages, undulating plateaux, dissected ridges

classification of more than 60 recurrent types of sandstone microforms. However, it follows from our introductory comments that classification is not one of our primary aims. Indeed, we turn briefly to it now only to avoid ambiguity in the following chapters. The brief classification outlined in Table 1.1 is based on variations in morphology, not on variations in a supposedly dominant control; and it assumes a hierarchical grouping of forms that can be considered at varying scales. A similar, though much more detailed approach to classification has been used with success for granite landforms by Twidale (1982).

In Chapter 2, we review the variations between and within masses of sandstone. We discuss the forms and processes that produce landforms at scales 2 and 3 (Chapters 3–7) and then we show how assemblages at scale 4 vary across climatic (Chapter 8) and tectonic (Chapter 9) regions. While the emphasis given in the following chapters to the variable properties of sandstone and their response to chemical and mechanical processes owes much to Yatsu's pioneering research in rock control, the use of regional studies attempts to extend Mainguet's admirable comparative analysis of sandstone terrain.

2

Variations within sandstones

The profound effect of differences in the properties of the rock on landforms in the Kimberley region of northwestern Australia emphasizes the importance of the diverse nature of sandstones, and the way in which they are spatially distributed. The properties of sandstones vary with the source of the clasts, the degree of weathering of the parent rock, sorting and abrasion during transportation, alteration following deposition, and the degree of recycling after initial deposition (Nesbitt *et al.*, 1997). Although these topics are dealt with in great detail in the sedimentological literature, some that are of particular significance to sandstone geomorphology need introductory comment.

Sedimentation of sand

Sand is usually defined as having a particle diameter of 0.063–2.0 mm (63–2000 μm). Particles finer than sand are silt, and then clay, clay particles having diameters finer than 0.002 mm (2 μm) or, in some classifications, 0.004 mm (4 μm). Coarser particles are gravels (usually with an upper limit of 20 mm or 64 mm). In the inverse log ø scale often used in sedimentological literature, sands are from −1–4 ø. While the definition of sand is based on particle size, not mineralogy, quartz sands dominate many sandstones throughout the world because of the resistance of quartz (relative to other minerals) to weathering and abrasion.

As Adamovic and Kidston (2008, p. 13) note:

Huge sandstone bodies typically develop in peripheral parts of rapidly subsiding basins adjacent to rising landmasses. Settings favourable for the formation of mineralogically mature sandstones and conglomerates include deserts and shallow seas in extensional/ transitional basins, and shallow seas on passive continental margins.

Sands are deposited in a wide range of sedimentary environments, and their characteristics reflect this range. Deposition in continental basins, during slow uplift of a nearby massif, allows time for feldspars and lithic fragments to be weathered, and predominantly quartzose sands deposited. Rapidly subsiding basins fed by rising mountain chains are more likely to have feldspar-rich sands; and along island arc collision zones, lithic fragments may dominate. Heterogeneous sediment is sorted and altered during large-scale deposition, such as occurs in sedimentary basins. Near a source area, sediments are likely to be coarse, with a relatively wide range of particle sizes (poorly sorted), with angular rather than rounded particles, and with heterogeneous mineralogy. In general, clays and silts are transported further than sands and gravels, creating facies changes across a mass of sediment.

On beaches and in streams, sand often moves as sheet-flow bed-load, that is, as a layer of sediment many times the thickness of the sediment diameter, moving along the bed. Finer sediment is more easily entrained and carried by turbulent flow as suspended load. Coarser load moves as initial-motion bed-load; some grains on the bed are dislodged and move some distance, whereas others stay stationary. In both environments, the bed-load may form ripples. Sand avalanching down the face of ripples is sorted, with finer sand being winnowed away in the turbulence and relatively coarse sand dropping at the base of the face. In ripples, in transport of sand down valley slopes, and in many other micro- or meso-environments, the sorting of sediment so familiar on a basin scale is echoed. Wind-blown sand commonly forms dunes along coasts, beside rivers and in desert areas. Particles finer than $100\,\mu m$ (0.1 mm) can be lifted in suspension and blown thousands of kilometres; clays are often aggregated into too massive a structure to be lifted by winds; and sands coarser than 2 mm are rarely lifted by fluid flow, but may move short distances by saltation. Sand dunes thus usually reflect the grain size characteristics and mineralogy of nearby source areas, and are well sorted.

The size of the constituent clasts in a sandstone depends on the rock from which the sand was derived, and thus on the resistance of the clasts to abrasion. But it also depends on the turbulence of the transporting medium and on the distance over which the sand has been moved. Generally, the closer the source, the larger are the clasts. A relatively high content of readily weatherable minerals is also usually the result of proximity to the source – granitic terrain for feldspars, and volcanic terrain for basaltic or andesitic clasts. A high content of quartz may be indicative of a quartz-rich source area, or of the destruction of other minerals during prolonged transport. Indeed, largely pure quartzose sandstones were long considered to be necessarily the result of repeated recycling, but Krynine (1941) suggested that highly quartzose sands could be produced from sediments derived from granitic rocks, without recycling, if chemical weathering during and after transportation was sufficiently intense and prolonged to destroy the feldspars. His contention has been sustained by the recognition of unambiguously first-cycle,

Fig. 2.1 'Beehives' in the Aztec Sandstone, The Valley of Fire, Nevada, USA show steeply dipping cross-beds

highly quartzose sands in the Orinoco basin of Venezuela and Colombia (Johnsson *et al.*, 1988). Rapid erosion and transportation from the granites in the Guyana Shield on the southern side of the basin results in feldspar-rich sediment, but weathering during slow transportation and temporary storage results in highly quartzose sands. Even the more complex lithology of the Andes on the northern side of the basin ultimately yields quartzose sands if weathering is sufficient to remove the more labile minerals. And this environment of intense chemical weathering and prolonged storage of sediment is not unique or even unusual.

Sedimentary structures preserved in sandstones give insight into the original mode of deposition, and also frequently control the micro-relief on sandstone outcrops and the movement of water through the sandstone mass (Adamovic and Kidston, 2008). Dunes built by aeolian transportation characteristically have large-scale cross-bedding inclined steeply (up to 35°) to the separation planes between beds, although flat-bedding may be developed in the swales (Fig. 2.1). Sand deposited by streams and waves tends to have alternating cross-beds and flat-beds, and to be inter-bedded with shales or mudstones. This is also the case with conglomerates, for while the pebbles are often approximately parallel to the separation planes of the beds, they can be inclined to the bedding. There are often also variable patterns of ripple-bedding and slumping. And sands deposited in stream channels or on offshore bars may occur as elongated stringers inter-bedded with finer sediments. Along shorelines, sands deposited between high and low tide levels are usually well sorted and rounded by surf abrasion. Between low tide level and the depth of reach of wave activity, sands are usually poorly sorted and cross-bedded. However, sands deposited by turbidity currents into very deep water beyond the continental shelf generally have considerable amounts of finer matrix, and display well-developed cycles of graded bedding due to the settling of coarser before finer sediment.

Changes after sedimentation

Compaction and diagenesis following deposition can cause major differences in the properties of sandstones. The muddy sandstones, or greywackes, of New Zealand are estimated to have been buried to depths > 15 km, and the great pressures exerted on them have produced a hard, but complexly jointed, rock (Coates, 2002). These sediments have also behaved in a semi-plastic manner during deposition and were crumpled into a complicated series of folds. In contrast, the remarkably quartzose St. Peter Sandstone of the north-central USA was deposited as an extensive sheet in a shallow marine environment, and has large-scale cross-stratification. Extensive quartz overgrowths, like those on the grains of this sandstone, are generally attributed to pressure solution at considerable depth, but oxygen isotope analysis indicates that they developed at temperatures ranging from $10\,°C$ to $40\,°C$, and are thus probably the result of near-surface dissolution of grain boundaries and inputs of silica from groundwater in a fashion akin to the formation of silcretes (Kelly *et al.*, 2007). Subsequent weathering and removal of cement by groundwater has in places left this sandstone almost cohesionless, but the abraded surface of the grains imparts high angles of friction (57–63°) that allow it to stand in high bluffs along many valley sides (Dittes, 2002). Variability of cementation is common, even in a single unit. For example, much of the Tuscarora Sandstone, a prominent ridge-forming rock in the Appalachian fold belt, displays a banded appearance of alternating well-cemented and poorly cemented zones. The presence of argillaceous material in some beds appears to have inhibited the development of secondary silica growths responsible for much of the cementation in this formation (Heald and Anderegg, 1960). Differences in the mineral composition of sands, together with the dissolved content of groundwater, can result in intergranular cements ranging from calcite, through various iron compounds to pure quartz.

As is well documented from oil reservoir studies, the porosity (the ratio of void volume to total volume of the rock) and permeability (the rate at which a fluid can move through the rock) of sandstones depend on the grain size and cementation of the rocks. Permeability is strongly influenced also by the geometry of the beds (tabular, lenticular, etc.), bedding structures (such as cross-beds) and intercalated beds (such as mudstones). Faults, jointing and fracturing can increase the macro-porosity. For example, the Sussex 'B' sandstone of Wyoming, which has yielded almost 25 million barrels of oil, shows clear evidence that its initial porosity at the time of sedimentation was 35–40% for trough and planar cross-bedded facies. Burial led to rearrangement of the grains; and pores were partly filled by pressure-dissolved quartz, clay, feldspar and iron hydroxide. The final porosity averages about 24%, but within the formation, there is significant variability. Fine- to medium-grained, cross-bedded sandstone has 10–22% porosity and permeability of > 12 mD (millidarcies); whereas fine-grained sandstone with appreciable mudstone or calcite-cemented beds has much lower porosity (1–15%) and permeability (0.1–15 mD) (US Geological Survey, undated).

Classification of sandstones

Geological classification

Sandstones can be most simply defined as detrital sedimentary rocks in which sand-sized fragments are dominant. Within this broad scope, there is a considerable variation in the dominant grain sizes and thence properties; for example, very fine sandstones differ in appearance and properties to coarse sandstones. The variability between sandstones is further increased by the amount of material finer or coarser than sand that is present, for a fine, silty sandstone is very different from a coarse, pebbly one. The proportion of intergranular matrix is also significant, and 'clean' sands or arenites (< 15% matrix) need to be distinguished from 'dirty' sands, known as wackes or greywackes (> 15% matrix) (Figs 2.2, 2.3). The boundary between wackes and mudrocks (pelites) is generally taken as 75% matrix. Although carbonate and other non-siliceous sands are conventionally excluded from consideration, the composition of the grains can vary greatly. Coastal aeolianites may have sufficient calcareous shell material to be more correctly described as limestones than as sandstones. Even within the arenites, distinction must be made between quartz (< 5% feldspar or rock fragments), sublithic (5–25% rock fragments, excluding feldspar), lithic (> 25% rock fragments, excluding feldspar), subarkose (5–25% feldspar), arkose (> 25% feldspar) and volcanic (> 50% volcanic fragments) subtypes (Pettijohn *et al.*, 1972).

The relative abundance of the various types of sandstones remains conjectural. Probably the most widely accepted estimates (Pettijohn *et al.*, 1972) are approximately 35% quartz arenite, 15% arkose, 25% lithic arenite and 20% wacke. However, Meybeck (1987) has suggested that almost 80% are quartzose, 5% arkose and 15% wacke. Although there would thus seem to be considerable justification for concentrating geomorphological research on the arenites, especially those which are quartzose, the limiting of investigation to these rocks (the *greseux* of recent French studies) would give a lopsided view of landforms developed on sandstones.

Fig. 2.2 Strongly cemented quartz sandstone, with horizontal bedding and strong vertical jointing

Fig. 2.3 Greywacke, showing quartz veins and minor folding

Although technically clastic rocks in which the dominant particles are $> 2\,\text{mm}$ in diameter, conglomerates are included here because they generally grade laterally into sandstones or are inter-bedded with them. Furthermore, many conglomerates have a sandy matrix. The admixture of sand and gravel in the same deposit prompts subdivision of these rocks into conglomerate ($> 50\%$ pebbles), sandy conglomerate (25–50% pebbles) and pebbly or conglomeratic sandstone ($< 25\%$ pebbles). Poorly sorted conglomerates generally contain much clay as well as sand in the matrix, and can thus be classified as greywacke conglomerate. The size of the dominant clasts also allows further subdivision into fine (pebble), medium (cobble) and coarse (boulder) conglomerates.

While this classification is widely used for general descriptive purposes, its basis is essentially sedimentological, rather than geomorphological. Attention should be given to variations in the amount and composition of the intergranular cement, the porosity and permeability, and the pattern of bedding and jointing, in particular examples of these types of sandstones.

Engineering classification

Engineering classifications of sandstones are based on their variable geomechanical properties. Considered in this way, sandstones range from tightly packed but uncemented sands, through variably indurated rocks that fracture around the individual grains, to quartzites that break both through the cement and the grains.

Strength is defined as the applied stress at which a given rock fails, and 'the centrepiece in rock mechanics testing is the unconfined or uniaxial strength (UCS) test, to which most properties are compared' (McNally and McQueen, 2000, p. 186). The compressive strengths of sandstones range very widely, from 2 MPa to > 200 MPa, and vary with mineral content, grain packing, cementation and capillary tension. However, caution is needed in interpreting published results of uniaxial testing, for sandstone is non-homogeneous as well as anisotropic in its geomechanical properties, and the standard deviation for UCS tests on sandstones is typically 30–40% of the mean. Moreover, the UCS is not simply an intrinsic property of the sandstone tested, but varies with specimen size, test procedures and other factors, especially with the degree of saturation of specimen. The wet strength is invariably less than the dry strength, the difference ranging from about 25–95% for sandstones in the Sydney Basin of Australia (McNally and McQueen, 2000). Similar care is needed in the interpretation of results of compressive strengths estimated in the field by the use of a Schmidt Hammer.

The tensile strength of sandstone is considerably less than the compressive strength, the ratio ranging from about 1:10 to 1:16 in the Sydney Basin (McNally and McQueen, 2000). Although the ratio depends considerably on cementation, quartz and chert are especially brittle, and thus have exceptionally low tensile strengths relative to their compressive strengths. Again, care is needed in interpreting the results of testing. The Brazilian Tensile Strength (BTS), which is determined by splitting a rock disc under diametrical compressive loading, may underestimate the true strength by a factor of 1.5. The Modulus of Rupture Test (USBM) is a flexural bending test that simulates failure conditions much better than the Brazilian, but requires a longer core (130 mm) and thus cannot be applied to weaker sandstones. The Point Load Test (PLSI) is the most widely used, but the UCS can often be $10–45 \times$PLSI, rather than the $24 \times$PLSI generally assumed. Both the PLSI and the BTS are only valid where tensile breakage actually occurs, and do not apply in clay-rich and laminated sandstones, where failure seems due to progressive crushing and shearing (McNally and McQueen, 2000).

The deformability of sandstones is stated in terms of the Elastic or Young's Modulus and the Poisson Ratio, both usually being measured by strain gauges glued to UCS cylinders. The modulus generally quoted is the slope of the middle linear portion of the stress/strain curve, referred to as the tangent modulus at 50% of the failure stress (Et_{50}), where deformation is most nearly elastic. The modulus ratio (E/UCS) is an index of stiffness/strength characteristics useful in predicting rock failure. The ratio for very weak or porous sandstones is below 100 when they are wet, 100–200 for weathered sandstones, 200–300 for many unweathered sandstones, 500 for strongly silicified sandstone and may be up to 1000 for strong quartzites (McNally and McQueen, 2000). Weak sandstones may deform by pore closure when stressed. Clay content is also important, as the stiffness of the rock will vary with moisture content. Strong sandstones with low porosity fail at higher stresses but exhibit smaller strain. Strongly indurated sandstones and quartzites shatter at very high stress but at very low strains, with negligible prior yielding.

The classification given in Table 2.1 also provides a useful guide for field interpretation, but caution is prompted because weathering of sandstones may be accompanied by substantial changes in geomechanical properties. McNally (1993) detailed such changes in the Terrigal Formation, which is predominantly quartz arenite or lithic arenite, with inter-bedded siltstone, located north of Sydney, Australia. There was a marked decrease in compressive strength from 100–60 MPa in fresh sandstone to 40–30 MPa in slightly weathered sandstone, and then a further decrease to about 25 MPa in moderately weathered sandstone. The point load tensile strength decreased from about 2 MPa in fresh sandstone to about 1 MPa in moderately weathered sandstone. The most dramatic change was in the Elastic Modulus, which decreased from about 6 GPa in fresh sandstone to 1 GPa in slightly weathered sandstone. In short, only slight weathering caused major changes in these properties. McNally's study showed also that the mass permeability of the Terrigal Formation sandstones was affected in two ways, one

Table 2.1 *Geomechanical classification of sandstones (after McNally and McQueen, 2000)*

Type of sandstone	Geomechanical characteristics
Very dense sands	Extremely low strength, estimated UCS <0.6 MPa. A largely compact granular material. Strength dependent on intergranular friction and confinement, and on a small component of capillary tension acting when partly wet but lost when saturated or dry.
Locked or friable sands (Class V Sandstone)	Very low strength, estimated UCS 0.6–2 MPa. Strength derived from interlocking grain fabric, minor quartz overgrowths and, in former glacial areas, overcompaction by ice.
Sand rock (Class IV Sandstones)	Low strength, estimated UCS 2–6 MPa. Sparsely cemented with grain-to-grain contact in an open framework. Capillary tension in clay matrix reduced by wetting. Highly weathered sandstones, and also humicretes bound by organics.
Weak sandstone (Class II and III)	Medium strength, UCS 6–20 MPa (0.5–20 MPa when wet). Failure mainly around the grains and through the matrix, and by grain dislodgement and rolling in uncemented zones.
Strong sandstone (Class I)	High strength, UCS 20–60 MPa (upper boundary arbitrary). Breakage both around grains and through cemented grain contacts.
Indurated sandstone	Very high strength, UCS 60–200 MPa. Strongly cemented, low porosity. Failure mainly through grains immobilized by cement. Includes many rocks otherwise described as quartzites.
Quartzites	Extremely high strength, UCS >200 MPa. Very hard rocks including silcretes and orthoquartzites strongly cemented by silica.

counteracting the other. Stress relief following erosion-induced fracturing increased the bulk porosity. However, as weathering progressed, clays and colloids infilled these spaces, impeding fluid movement.

The effects on landforms of variations in sandstones

Examples from England

The effects on landforms of variations in the properties of sandstones have been demonstrated clearly in the study of four sandstones at inland sites in England by Robinson and Williams (1994).

- The Fell Sandstone outcrops as an impressive west-facing, crag-topped escarpment in the extreme northeast of England. This 330-m thick formation consists dominantly of

Fig. 2.4 Subdued cliffs on the Millstone Grit, west side of the Pennines, central England

massive cross-bedded sandstones, with occasional pebbly beds. It is quartz-rich, with only about 10% feldspar and 5% mica, has a moderately low porosity (9.4–14%) and is strong (UCS 74 MPa dry and 52 MPa wet).

- The Millstone Grit sandstones of central England alternate with argillaceous rocks, especially lower in the sequence. There are marked lateral variations in the thickness of these sandstones, and some are present for only short distances before grading laterally into shales. Some of them are fine-grained and flaggy, but most are medium- to coarse-grained. Many are felspathic (up to 27.5%), but porosities are generally low (7–17%). Strengths are relatively high when dry (39–104 MPa), but much lower when wet (24 MPa). They are relatively resistant to weathering, they often cap escarpments, and they outcrop as cliffs and crags with isolated pinnacles and tor-like masses (Fig. 2.4).
- The Ardingly and Ashdown Sandstones crop out as discontinuous lines of valley-side crags and cliffs in the Weald of southeastern England. These sandstones are poorly cemented, porous (26–27%), and are of moderate strength (31–51 MPa dry, 13–51 MPa wet).
- The New Red Sandstone forms discontinuous rounded to sub-rounded cliffs along the Welsh border. It is poorly cemented, relatively porous (8.9–25%) and is of relatively low strength (11.6 MPa dry, 4.8 MPa wet).

Clearly, the change from prominent cliffs to rounded hills is dependent mainly on the strength of these sandstones and their resistance to weathering. Robinson and Williams (1994) also noted significant differences between the morphologies developed on much weaker sandstones in England. Even weakly cemented sands and sandstones, such as the Folkstone Beds, can form locally higher ground, seemingly because their greater porosity restricts runoff and erosion. However, clay-rich or silty sandstones, such as the Sandgate Beds of the Lower Greensand, are not only weakly cemented, but have low infiltration rates, and are thus very susceptible to weathering and erosion. Similar relationships can be demonstrated for different types of sandstones worldwide.

Examples from southeastern Australia

The Hawkesbury Sandstone of the Sydney Basin is dominantly quartz arenite to quartz-rich sublitharenite. Most of it can be classified as strong sandstone, but some heavily weathered outcrops are more correctly classified as weak sandstones or sand rocks. About 95% is quartzose sandstone, and the remainder consists of sandstone laminite, with siltstone and claystone interbeds. There are prominent silica overgrowths on the quartz grains; the intergranular cement consists mainly of secondary quartz and siderite; and the argillaceous matrix consists mainly of illite and kaolinite (Bowman, 1974). Uniaxial compressive strengths (dry) generally range from about 35–55 MPa, and in iron-cemented bands, up to 125 MPa; the strength of wet specimens is generally from about 0.3–0.8 of the dry strength; the compressive strength is generally 12–20 times greater than the tensile strength; the elastic modulus is from about 2.5–16 GPa; and the porosity increases with the degree of weathering from about 13–20% (Pells, 1985; McNally and McQueen, 2000).

The Hawkesbury Sandstone seems to have been laid down by a very large river system, perhaps comparable to the modern Brahmaputra (Conaghan, 1980). It displays the usual depositional styles of braided streams – which Conaghan has designated as sheet facies – but also contains massive facies. The frequent planar and trough cross-stratification of the sheet facies exerts a major control on the micro-topography of benches and other gently inclined slopes. Cross-beds also control much of the indentation due to weathering on cliff faces, and provide preferential paths for small- to medium-scale fracturing (Fig. 2.5). However, where the cross-bedding occurs in the very thick sets (4–8 m) typical of the Hawkesbury Sandstone, jointing and bedding planes largely control the geometry of outcrop. Massive sandstone – that is structureless or faintly parallel-stratified sandstone (Jones and Rust, 1983) – is relatively uncommon in fluvial deposits, but occurs frequently within the Hawkesbury Sandstone, generally filling depressions perpendicular to the direction of flow. Jones and Rust attributed these massive bodies to deformation during deposition, suggesting that liquified sand, including laminated muddy intraclasts, failed down foreset slopes in erosional depressions at stages of falling river levels. As the muddy intraclasts generally weather more rapidly, they tend to form irregular recesses in the outcrop of the

Fig. 2.5 Cross-bedded Hawkesbury Sandstone. An infilled channel can be seen near the base of the cliff

massive facies. Conaghan has noted that, at least in coastal localities, the massive facies seem less resistant than the sheet facies, being often the sites of recessed sections of cliff faces. As outcrops of massive facies along coastal cliffs also appear to be preferential sites for the development of cavernous weathering, the topographic contrasts between the two facies may reflect varying susceptibility to chemical breakdown as well as differing mechanical properties.

The terrain for some 20 km or so south of Sydney is cut almost entirely in the Hawkesbury Sandstone, which rises very gradually in a southerly direction. The upland surface is composed of broad, stepped surfaces that descend gradually, then drop abruptly into narrow valleys. Many of these surfaces appear to have developed along the major bedding planes separating cross-bedded units from more massive, flat-bedded units. Seepage along the bedding planes causes the flat-bedded units to retreat, exposing extensive platforms on the cross-bedded units. The platform surfaces subsequently break down as runoff from further upslope penetrates into the exposed tops of the cross-beds that provide innumerable pathways for seepage. Many of the flat-bedded units have been eroded into elongated, sub-parallel ribs with rounded surfaces, known locally as 'whalebacks' (Fig. 2.6). These features are the result of the deepening and widening of the surface exposures of the dominant north–south jointing.

Fig. 2.6 'Whalebacks' on the Hawkesbury Sandstone plateau south of Sydney, Australia. Note the cavernous weathering of the face at the end of the joint block

Weathering of sandstone on the plateau is aided by the development of extensive swampy heath and sedgelands that retard runoff and facilitate continued seepage long after rainfall has ceased. The main valleys are generally less than 1 km wide, and in places are only a few hundred metres across. As there is little basal sapping near the floors of the valleys, variations in the strength of the sandstones, especially where clayey interbeds occur, are dominant, and have produced a series of stepped slopes down the flanks of the valleys, rather than a single major cliffline. But major clifflines do occur where beds are particularly massive and where there are no significant clayey interbeds. Basal sapping beneath the sandstone, and the consequent development of a major cliffline, occur further south where claystones outcrop beneath the Hawkesbury Sandstone, and along the adjacent coast where erosion by waves has produced spectacular cliffs. Similar relationships occur in the Blue Mountains, west of Sydney. In the western part, where claystones and siltstones are exposed beneath the sandstone, high cliffs fringe wide valleys; whereas in the east, only sandstone is exposed, and stepped cliffs flank narrow valleys.

The Budgong Sandstone, which outcrops further south in the Sydney Basin, is a volcanic arenite, with inter-bedded laminated siltstone. The dominant sand-sized material is basaltic and andesitic detritus, and the pebble and cobble component is of the same origin (Bowman, 1974). The matrix (mainly of illite, degraded illite and kaolinite) is only a minor component of the rock. As 30–40% of the volcanic detritus is plagioclase, the sandstone weathers fairly readily. Weathering extends 1–3 m below the surface, and to depths of up to 10 m down major joints. Therefore, although the Budgong Sandstone stands in vertical cliffs along the coastline where it is undercut by waves, it generally forms much gentler slopes of 5–20° on hillsides. Benching occurs where the sandstone is inter-bedded with several major latite flows. However, the contrasting permeability of the sandstone and latite makes these sites very prone to mass slumping. Away from the coastal source of the volcanic debris, the Budgong Sandstone becomes increasingly siliciclastic, and is more properly classified as a lithic arenite. This change in composition and in susceptibility to weathering has a pronounced effect on rock strength, and the sandstone inland forms prominent cliffs on the midslopes of the valleys.

Examples in southwestern and central USA

Although the Navajo Sandstone and the Entrada Sandstone of the Colorado Plateau have a high feldspar content, they consist of > 80% quartz sand (Howard and Kochel, 1988). They are therefore subarkosic arenites rather than true arkoses. The grains are weakly cemented by calcite, hematite and, less frequently, by silica. As their UCS ranges from about 17–50 MPa (Schumm and Chorley, 1966), they can be classified as weak to strong sandstones. They are certainly strong enough to stand in high vertical faces, such as the Great White Throne cut in the Navajo Sandstone in southern Utah (Fig. 2.7). However, the major clifflines

Fig. 2.7 High vertical cliffs with strong jointing, on the Great White Throne in the Navajo Sandstone, Zion National Park, USA

of the Colorado Plateau occur where the resistant caprock is undercut, either by the breakdown of weaker or more closely jointed rock (Howard and Kochel, 1988; Nicholas and Dixon, 1986), or by groundwater sapping. The porosity of the Navajo and Entrada is generally 20–35%, and the permeability $2 \times 10^{-2}-1$ darcy, with the Entrada at the lower end and the Navajo at the upper end of the range (Howard and Kochel, 1988). Variations in total permeability are controlled also by jointing and by the dominant characteristics of the bedding, especially by the contacts between the abundant large-scale cross-beds, and the horizontally bedded sandstones and mudstones that were deposited in interdune swales of these aeolian sandstones. The dominant movement of groundwater down the dip of the Navajo Sandstone has resulted in large-scale basal sapping that has been a major control on the extension and morphology of many canyons (Laity and Malin, 1985; Laity, 1988). Another outstanding feature of this sandstone terrain is the extensive development of generally low rolling relief on bare rock surfaces. Erosion of these 'slickrock slopes' is dependent on the interaction of seepage and runoff, exfoliation, jointing and major bedding partings (Oberlander, 1977; Nicholas and Dixon, 1986). Slickrock slopes, together with groundwater sapping and other solutional features, are considered in more detail in later chapters.

The importance of relatively small-scale variations in the character and properties of these arkosic sandstones is illustrated dramatically by the giant weathering pits on the summits of conical hills in the Entrada Sandstone on the northern side of Lake Powell in southeastern Utah (Fig. 2.8). The conical features recorded by

Fig. 2.8 Conical hills eroded from Entrada sandstone. Small pits etch the bedding planes. (Photo: D. Netoff)

Fig. 2.9 A huge weathering pit (15 m deep) on a slickrock slope in the Entrada Sandstone. For scale, note the people. (Photo: D. Netoff)

Netoff and Shroba (2001) in this area range from low domes with < 5 m relief and < 10° slopes, to necks with > 35 m relief and slopes from 45–90°. The maximum local relief is 75 m. Many of these conical features have no pits on their summits, but others have pits from < 3–70 m wide and 2–16 m deep. The flanks of the conical hills are composed mainly of cross-bedded sandstone, whereas their summits and summit weathering pits formed in clastic pipes of massive sandstone and breccia (Fig. 2.9). The pipes are cylindrical masses that appear to have been injected in a fluidized state into the cross-bedded host rock. Many of the pipes are bounded by ring faults; and where there is clear evidence of movement along the faults, the pipe material has been displaced downwards by 2–5 m, possibly owing to the compaction of pipe sands during emplacement. Erosion of the friable host rock from the flanks of the hills has commonly exposed near-vertical walls of pipe material. The preservation of the cones seems to be the result of the variable content of calcite cement. In four transects, the cement content ranged from 8.2–22.9% in the contacts between the pipes and the host rock, 4.5–6.5% in the core of the pipes, to 5.3–6.9% in the host rock. Although the sandstone decays by a series of weathering processes, abrasional features in the pits indicate that the weathered debris has been removed by wind in this arid climate, rather than by the solutional processes observed in more humid areas.

Examples in Norway

The *sparagmite* sandstones of central-eastern Norway are arkosic. Some have a feldspar content up to 30%, and some contain orthoclase grains as large as 4–5 mm (Holtedal, 1960; Strand and Kulling, 1972). However, the landforms developed on them are very different from those on the arkosic sandstones of the Colorado Plateau. This difference is not only a result of the shaping of the *sparagmite* terrain by glacial, periglacial and fluvial processes; it also reflects differences in the internal characteristics of the sandstones. The *sparagmite* sandstones are largely of shallow marine origin, and some of the conglomeratic units in them are glacial tillites. Thus, instead of the large-scale cross-bedding typical of the Navajo and Entrada Sandstones, the finer-grained beds of the *sparagmites* are typically flat-bedded and have outcrops that are typically flaggy. In the Rondane Mountains, they have a uniform northerly dip over considerable distances (Strand and Kulling, 1972) (Fig. 2.10). Although there are steep slopes on the walls of relict cirques in the Rondane Mountains and on valley sides, these sandstones do not form high continuous cliffs, and do not have large

Fig. 2.10 Rounded summits with relict cirques cut in *sparagmites* of the Rondane Mts, Norway

groundwater-sapping features. And instead of slickrock slopes, these flaggy sandstones have been eroded into broad, gently sloping surfaces between the mainly conical hills that rise above them. Slopes tend to be stepped along major bedding features, although some of the more extensive benches or basins appear to be relict erosional surfaces.

Landforms on quartzites

The Sioux Quartzite, which outcrops widely in the north-central United States, is an indurated quartz arenite of Proterozoic age (Southwick *et al.*, 1986). It was deposited in braided channnels, and the basal section is conglomeratic. The grains in the sandstone are dominantly monocrystalline quartz, and are strongly cemented with silica. There are both flat-beds and cross-beds, and joints are well developed. Where the rock is incised by streams, the combination of prominent bedding and jointing has resulted in canyon walls consisting of alternating benches and vertical faces. Elsewhere, outcrops generally have low relief, up to 5 m, with joints opened by weathering in fluted and polished surfaces.

Landforms eroded in quartzite depend not only on bedding and jointing, as discussed for the Sioux Quartzite, but also on the thickness of the rock and the fracture patterns developed in it. At Zhangjiang, in the Wuling Shan of western Hunan in central China, thick quartzite with strong vertical jointing has been eroded into an amazing 'rock forest' of numerous towers, many of which are 100–200 m high (Zuo and Xing, 1992, p. 190). In marked contrast, the highly quartzose silcretes that cover large areas of Australia and southern Africa, and also occur in parts of North America and Europe, generally form a discontinuous tabular terrain (Fig. 2.11). Although their high quartz content (> 90%) makes them resistant to weathering, they are rarely more than 2–3 m thick, most having formed in siliceous sediments on ancient valley floors. The silcrete is left as a thin caprock by erosion of the much weaker, generally weathered rock beneath.

Greywackes in New Zealand

The poorly sorted sandstone, known as greywacke (from the German *grauwacke*), makes up much of the landmass of New Zealand. Sediment that had accumulated on the eastern continental shelf of Gondwanaland was carried by turbidity currents, which maintained the mix of coarse and fine particles, into much deeper water. Deposition of much finer sediment in the deep marine environment resulted in these sandstones being inter-bedded with argillaceous material. Subsequent deep

Fig. 2.11 A deep silcrete crust in central Queensland

burial (Coates, 2002) produced hard, though very closely jointed, sandstone. Unweathered greywacke is typically strong to very strong, having unconfined compressive strengths ranging from about 90–350 MPa (Read *et al.*, 2000). Because the greywacke is generally intensely deformed, however, defects in the rock mass are closely spaced. Average spacings range from 60–200 mm; spacings greater than 750 mm are rare; and spacings < 60 mm are common in crushed and sheared zones. Thus, notwithstanding the high compressive strengths of individual blocks, much greywacke has only a fair to poor rock mass strength.

Folding of the greywacke during the Late Mesozoic, followed by intense weathering to depths as great as 90 m during Tertiary times, further increased the variability of its strength and resistance to abrasion. Erosion following elevation of this rock mass by faulting in the Late Tertiary has produced an intensely dissected terrain, consisting mainly of closely spaced V-shaped valleys and long narrow ridges (Suggate, 1982) (Fig. 2.12). While most of the greywacke of New Zealand is

Fig. 2.12 Greywacke boulders eroded from the Southern Alps along the channel of the Rakaia River, New Zealand. Note the intensely dissected ranges

steeply dipping, the gently inclined strata of the Lord Range (Andrews *et al.*, 1974), which lies near the main divide of the Southern Alps, give scope for comparison with the other generally gently dipping sandstones considered here. The rock sequence of the Lord Range is extremely uniform – sandstone is everywhere thicker than siltstone, and dominant in most places, except on the western side of the range where there has been metamorphic alteration of the greywacke to schist. The gently dipping and massive sandstones have been shaped into a castellated, cliff-and-bench topography, with steep faces up to 450 m high.

Landforms on conglomerates and breccias

Although the strength of conglomerates and breccias varies considerably with the nature of the matrix and the cement, the strength of the rock is greatly increased where the clasts are in contact with each other and form an essentially continuous framework or fabric. The Meteora Conglomerate of central Greece consists of wedge-shaped masses of deltaic deposits with large cross-set bedding, into which were cut deep channels containing giant bars of cobbles and

Fig. 2.13 Strong joint control of steep caverned cliffs cut in conglomerates and sandstones at Meteora, Greece

sand (Ori and Roveri, 1987) (Fig. 2.13). This mass of conglomerate has been dissected into steep-sided towers and buttes that rise abruptly from the surrounding lowland. Much of the dissection has been located along steeply dipping joints. Many joints have also provided planes down which massive slab failure has occurred. Numerous caverns have developed on the middle to lower slopes of this very porous rock.

The spectacular landforms at Montserrat, in the Catalan Range of northeastern Spain, illustrate the importance of jointing in the shaping of conglomeratic masses (see, for example, Louis and Fischer, 1979, Bild 37). Weathering has extended preferentially down the strongly developed, vertically dipping, E–W and N–S intersecting joint systems through this horizontally bedded rock mass. The topography now consists of a tiered array of towers and pillars with rounded crests that are similar in appearance to those found in crystalline rocks. The platforms separating the tiers are developed along beds of sandstone.

Basin characteristics

In addition to the repetitive patterns of lithic content, fabric and structure discussed above, there are also much larger and systematic changes across sedimentary basins that have major effects on landforms. The basic factors that control the stratal architecture of basins are the sediment flux, changes in the capacity of the basin caused by eustatic or tectonic changes, and the general form or physiography of the basin (Posamentier and Allen, 1993). Basin physiography exerts a particularly important control by determining, for example, whether or not deep-water turbidites will be a major component, or to what extent the sedimentary sequence will be disrupted by erosion.

Judging from most textbooks on landforms, the characteristics of basins are of no great importance to the geomorphologist, and can be left to the stratigrapher. This is a short-sighted view. Consider the consequences for variations in the style and especially in the distribution of sandstone landforms implicit in the following brief summaries of Beloussov's (1980) description of the development of basins in eastern Europe. First, he noted a great contrast between the very thick, dominantly argillaceous and carbonaceous deposits of the Caucasian geosyncline, and the much thinner, more sandy sediments of the Russian platform. Because of these contrasting depositional characteristics alone, the extent and style of sandstone landforms in former geosynclines like the Caucasus differ from those of the platforms. Then, following Ronov, he documented the great variation in time and space of sand deposition over the Russian platform:

- Dominantly sandy deposition in the Cambrian was followed by overwhelmingly argillaceous and carbonaceous deposition during the Ordovician.
- An important phase of sandy deposition during the Devonian gave way to carbonaceous deposition in the Carboniferous and early Permian.
- This was in turn followed by a prolonged period from the Permian to the Early Tertiary, when sands comprised from 35–50% of sediments laid down on the platform.
- The main areas of sand accumulation shifted from the northeastern part of the platform during the Carboniferous, to the eastern part during the Permian, the southeastern part during the Upper Jurassic, and then to the central part during the Upper Cretaceous.

Thus the distribution of the types of sandstone, and thence the landforms developed on those rocks, is intimately related to changes in sedimentation in different geological periods.

In contrast to the diversity within and between the large basins described by Beloussov, the small Hornelen Basin between Nordfjord and Sognefjord in western Norway has been filled with remarkably uniform sandstones (Holtedahl, 1960; Steel, 1976). There is an enormous depth of fill, up to 25 km thick, notwithstanding a small basin area of $< 2000\,km^2$. What is more, and despite the great depth of fill, sediments finer than fine-grained sandstones are remarkably scarce. The entire basin is dominated by stacked alluvial plain sediments deposited in coarsening-upward

Fig. 2.14 Deep parallel gullies have been carved into the lower slopes of the Nuussuaq Peninsular, western Greenland

cycles. The adjacent Solund Basin is of similar size to the Hornelen Basin and also dominated by coarsening-upward alluvial cycles; but, whereas the latter is dominated by sandstone, the former is virtually filled with conglomerate (Steel, 1976). These contrasts in style of deposition are, as we shall see, reflected in contrasting topographies.

The Paleocene onshore sediments of the Nuussuaq Basin of Western Greenland indicate a sequence of uplift and valley incision, followed by rapid subsidence, that appears to have been a response to the development of a mantle plume prior to the widespread extrusion of basalt (Dam *et al.*, 1998). Uplift led to incision into Late Cretaceous mudstone and conglomerate. An extremely uniform stacking pattern of uniform sandstone, that lacks internal fining-upward successions, indicates that deposition in the valleys was very rapid, and was the result of rapid subsidence that increased accommodation space, and high rates of discharge and sedimentation. Glaciation, and in warmer periods periglacial action and deep gullying, have sculpted smooth slopes cut by parallel streams through this uniform lithology (Fig. 2.14).

The Chaco Basin of eastern Bolivia contains a 7.5-km thick wedge that consists mainly of coarse sediment transported from the eastern Cordillera of the Andes (Uba *et al.*, 2005). These sediments were deposited in a series of megafans, similar to modern fans at the front of the cordillera, that record the effects of uplift throughout Neogene times (Horton and De Celles, 2001). As the fans formed ahead of the eastward-propagating fold–thrust belt of the cordillera, the older sediments have deformed more than the younger sediments further out in the basin. Subsequent erosion, especially along the Comargo Syncline that lies 200 km west of the modern fans, has cut a rugged landscape in dominantly cobble and boulder conglomerate and sandstone.

The types of basins considered here are, in the classification of Fischer (1975), secondary basins, rather than the primary, or major ocean basins opened by rifting. These secondary basins result from geological modification of primary basins or of continental platforms. For example, marginal oceanic, or inter-arc basins, probably resulted from the foundering of oceanic lithosphere; rift basins on the continents resulted from tectonic stretching expressed in complex faulting and subsidence patterns; the deep basins of the interiors of the continents, that is the auto-geosynclines, may well result from deep-seated controls such as phase changes or mantle differentiation (Fischer, 1975). Moreover, as Fischer demonstrated from the Appalachians, basins may undergo profound changes. The Appalachian Valley and Ridge Province began as a basin of the trailing-continental type, was transformed to an orogenic condition during the Taconic episode, then changed to an auto-geosynclinal type, and returned to the orogenic state during the Appalachian episode. Regardless of tectonic setting, most basins also bear a strong imprint of eustatic change and of loading by water and sediment. These general considerations need to be kept in mind as we turn to the geomorphological effects of various aspects of the deposition of sandstone.

The influence of vertical sequences within basins

The Colorado Plateau, USA

It is obvious that variations in the source and rate of supply of sediments, together with fluctuations in the geometry of a basin caused by tectonic or eustatic effects, will be imprinted in the vertical sequence of sediments from which landforms are subsequently cut. Although the effects of such changes on individual rock faces have frequently been described, there have been surprisingly few attempts to systematically compare the geomorphological effects of contrasting vertical sequences. Oberlander (1989) has pointed the way with good illustrations of such effects on a selection of sandstone scarps on the Colorado Plateau:

- Where massive sandstone is by far the dominant outcrop, a thin substrate may have very little topographic representation.
- A massive caprock over a thick exposure of thinly bedded sandstone and shale substrate results in active retreat of the cliffs formed from the caprock and in extensive talus accumulation that retards dissection of the lower slopes.

- A thin caprock over highly erodible substrate also leads to active retreat of the caprock, but yields insufficient talus to retard dissection of the lower slopes.
- Very thin caprock may result in the dissection of almost the entire slope into the forms typical of badlands. However, in some cases, cliff segments on very thin caprock may extend down into the easily erodible substrate, producing steep substrate faces that dominate the scarp.

Oberlander's examples can, by and large, be considered as instances of the effects of the contrasting properties of different formations. There are also important, if more subtle, effects of vertical changes within individual formations. Consider firstly the effects of the rhythmic repetition of distinctive beds, as exemplified by the Hermosa Formation of the southeastern Colorado Plateau. Lithological repetition in this formation, and in the overlying Rico and Goodrich sediments, has given rise to the intricately stepped walls of the famed 'gooseneck' bends of the highly sinuous San Juan canyon (Gregory, 1938). Contrast this with the apparently similar repetition of thin beds of sandstone and gypsiferous mudstone in the Summerville Formation in eastern Arizona. Rather than having stepped slopes, this formation stands in vertical faces (Harshbarger *et al.*, 1957, Figs 27, 28).

The Hornelen Basin, Norway

Strong control of topography by bedding may develop even where sandstones are infrequently split by silty members, as is the case in the Hornelen Basin. The landforms now developed in the Devonian sandstones of this basin comprise a superb set of stacked and tilted benches extending for about 60 km (see Holtedahl, 1960, especially Fig. 94). Stratigraphic sections presented by Steel (1976) show that this benched topography is controlled overwhelmingly by facies variations within coarsening-upward cycles of alluvial sandstones. Some of the benches seem to have developed by preferential erosion along thin mud and silt laminae. Others are developed entirely within sandstone, apparently by the preferential erosion of plane-parallel laminae and of ripple laminae.

Benching is less prominent in the very thick accumulations of coarse conglomerate in the nearby Solund Basin (Holtedahl, 1960; Strand and Kulling, 1972). Although the basal section contains very coarse breccia in which individual blocks > 1 m were derived from adjacent granite and quartzite, and coarse sandstones occur higher in the sequence, coarse conglomerates dominate. Thrust faulting resulted in the shearing and tectonic deformation of parts of the conglomerate. Subsequent erosion has produced a bare and rugged terrain, in which the tiered slopes, as on many conglomerates, have been partly rounded by weathering and displacement of individual clasts (Fig. 2.15).

Fig. 2.15 Tilted beds of conglomerates and sandstones form impressive cliffs in the Solund Basin, western Norway

The Sydney Basin, southeastern Australia

The effect of vertical changes in resistance of sandstones to erosion is strikingly illustrated by the array of cliffs and benches carved from Paleozoic sediments in the Clyde Valley at the southern extremity of the Sydney Basin (Fig. 2.16).

- The lower slopes of the valley have been cut from folded slates and sandstones, above which there is a series of small benches, about 50 m thick, cut from the inter-bedded lithic sandstones and shales of the Pigeon House Siltstone. These soft, clayey, quartz felspathic sandstones form low bluffs 1–6 m high (Gostin and Herbert, 1973).
- Much of this unit is mantled by cobbles that have fallen from the overlying Yadboro Conglomerate. The cobbles and boulders in this conglomerate are up to 38 cm in diameter, and form an essentially continuous framework, the interstices of which are filled with coarse to medium quartz sandstone (Gostin and Herbert, 1973). The conglomeratic beds dip only at about 2°, and are cut by widely spaced and almost vertically dipping joints. Because of its considerable strength, the conglomerate forms massive cliffs. The main process of cliff retreat is the sapping caused by the growth of large caverns at the base of this very permeable rock mass.

Fig. 2.16 Sandstone cliffs in the Clyde Valley, southern Sydney Basin. The mesa is known locally as The Castle. The upper cliffline is Nowra Sandstone; the forested slopes are on Wandrawandian Siltstone; the lower bench and cliffs are in the Snapper Point sandstones and the Yadboro Conglomerate; the lower forested slopes are on folded strata

- The top 20 m or so of this 160-m high line of cliffs are cut from the Snapper Point Formation which lies disconformably on the conglomerate. As this formation consists mainly of pebbly, silt-free sandstone, it too stands in vertical faces.
- Above it is the Wandrawandian Siltstone, which consists of fine-grained quartz-lithic sandstone and siltstone (Gostin and Herbert, 1973). The lower part of this unit has been stripped from the surface of the Snapper Point Formation, which forms a distinct bench and is most prominent on the summit of the Byangee Walls mesa that is only 300 m wide. The slopes cut in the Wandrawandian are inclined at about 28°, and are in part mantled by massive blocks fallen from the upper cliffline cut in the Nowra Sandstone.
- The Nowra Sandstone is a dominantly flat-bedded quartz arenite, which is cut by steeply dipping joints and stands in vertical faces up to 170 m high (Le Roux and Jones, 1994).
- The flat-bedded quartz-lithic sandstones of the Berry Formation have been stripped from the top of much of the Nowra Sandstone, but remnants of it form a series of stepped benches and low cliffs 5–20 m high, above the prominent Nowra Sandstone clifflines.

Lateral changes across basins

Lateral variations in slope forms that reflect facies changes have been even more neglected than have the effects of vertical stratigraphic sequences. Even those sites long regarded as classic illustrations of rock control on slope morphology – like the view north across the Grand Canyon of the Colorado to Bright Angel Creek – must be recognized as part of a sequence whose lithologic and topographic characteristics change markedly over short distances.

The Colorado Plateau

The Kaibab Formation, which forms the prominent upper cliffs of the Grand Canyon, grades westward from a non-calcareous sandstone to a calcareous sandstone, thence to a sandy limestone and finally to a crystalline limestone, over a distance of about 45 km (McKee and Resser, 1945). Major changes in composition occur in other formations exposed along the canyon's walls, as for example in the Bright Angel Shale and the Muav Limestone, which become dominated by sandstone in the big bend area of the eastern Grand Canyon (McKee and Resser, 1945). There are also major changes in the thickness of formations, like the cliff-forming Coconino Sandstone (Fig. 2.17), which increases from 10 m in the Marble Canyon to 200 m in the eastern Grand Canyon, but then decreases to a feather-edge in the western Grand Canyon. Lateral changes such as these can have a major influence on regional landform variation, especially where dominant cliff-formers like the Coconino Sandstone pinch out.

Similar variations occur in prominent cliff-forming sandstones elsewhere on the Colorado Plateau (Harshbarger *et al.*, 1957). The Wingate Sandstone is commonly known for its bold, cliffed outcrops, but the Rock Point Member of this formation is carved into smooth slopes because of its high silt content. At its type locality, the Kayenta Formation is a sandstone which forms rounded slopes and cliffs, but 100 km to the southwest, it is represented by a dominantly silty facies that is dissected into an irregular badland topography. The Entrada Sandstone consists of two facies – a clean sandy one which forms massive cliffs; and a red silty sandstone characterized by spheroidal weathering and the development of weirdly shaped, 'hoodoo' outcrops. Even the Navajo Sandstone, which is the most lithologically uniform of the sandstones of the plateau, contains lenticular beds of cherty limestone that form conspicuous ledges in the southern part of the region.

The Wasatch Plateau, Utah, USA

These examples involve largely gradational changes across single formations. Facies changes often result also in an interfingering between formations. The

Fig. 2.17 High cliffs on the Coconino Sandstone, northern edge of the Grand Canyon, Colorado Plateau, USA

topographic effects of a very complex interfingering of sandstones and shales are illustrated exceptionally well along the escarpments of the Book Cliffs and Wasatch Plateau in eastern Utah. The alternating advance and recession of the Cretaceous shoreline resulted in successive accumulations of beach sand extending over fine sediments. These sands then pinched out, and were subsequently buried by fine sediments as the sea rose again. This sequence can now be seen as a series of tongues of sandstone that split, dip down eastward into, and finally pinch out in, the Mancos Shale, giving a complex zig-zag pattern to the facies boundaries (Spicker, 1949). Young (1955) identified no fewer than 14 prominent tongues of sandstones extending eastward, for variable distances, into the Mancos Shale (see also Fisher *et al.*, 1960). These changes produce a complex relationship between the rocks and the landforms along the Book Cliffs that can only be deciphered because of the exceptionally clear and continuous exposures (Fig. 2.18).

As the sandstones capping the Mancos Shale along the 350 km of the Book Cliffs have a consistent appearance and are visually striking, they readily give the impression of extensive continuity, and this was long thought to be so (Spicker, 1949). In fact, individual sandstones change their position on the cliffs and eventually pinch out. Their topographic expression is also greatly constrained by the thickness and composition of the rocks above and below them. The stratigraphic detail and excellent photographic illustration of these changes in an easterly direction along the cliffs presented by Spicker (1949), Young (1955) and Fisher *et al.* (1960) are the bases for the following description.

- In the west, on the eastern edge of the Wasatch Plateau, a bold escarpment consists of an upper cliffline of Castlegate Sandstone overlying about 300 m of the Blackhawk Formation, below which there is a secondary cliffline in the Star Point Sandstone and then lower slopes cut in the Mancos Shale. Sediments higher in the sequence here have been pushed back by stripping of limestones and shales from above the Castlegate.
- About 45 km to the east, near Horse Creek Canyon, the escarpment is more complex, owing mainly to the thickening of the sandstones of the Bluecastle Member of the Price River Formation between the Castlegate Sandstone and the overlying shales and limestone. The Bluecastle Member here forms the upper cliffline, which is separated from the Castlegate cliffline by slopes cut across shales. Below the Castlegate are secondary cliffs and shaley slopes in the Blackhawk Formation, then dissected footslopes in the Mancos Shale.
- Further east, near the Green River, the lithologic and topographic sequences seem similar, but important changes have occurred. Just to the west of the Green River, the Castlegate Sandstone is capped by the debris slopes and cliffs of the Bluecastle Member sandstones, without any prominent benching. Only 5 km to the east of that

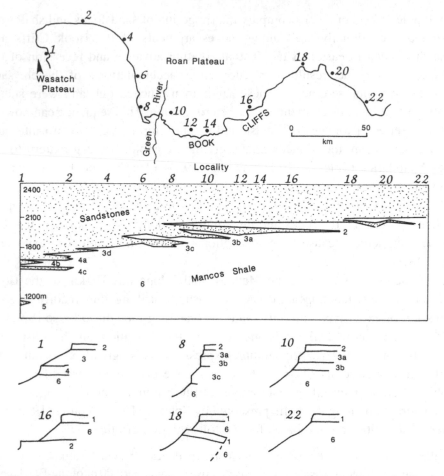

Fig. 2.18 Facies changes and variations in scarp morphology along the Book Cliffs, Utah (after Spicker, 1949; Young, 1955; Fisher *et al.*, 1960). Note the complex inter-fingering of sandstones and shales.

1 Sego Sandstone; 2 Castlegate Sandstone; 3 Blackhawk Formation: 3a upper sandstone member; 3b middle sandstone member; 3c lower sandstone member; 3d Aberdeen Sandstone member; 4 Star Point Sandstone: 4a Spring Canyon tongue; 4b Storrs tongue; 4c Panther tongue; 5 Emery Sandstone member of the Mancos Shale; 6 Mancos Shale (shale facies)

river, the Bluecastle sandstones have been pushed back, leaving a very broad bench at the top of the Castlegate Sandstone. This is because of the appearance and rapid thickening of the Buck Tongue of the Mancos Shale above the Castlegate. Below the Castlegate cliffs, the Blackhawk Formation is only about 50 m thick, and the lower cliffline above the Mancos Shale is no longer in the Star Point Sandstone, which has pinched out, but in another tongue of sandstone about 240 m above the stratigraphic horizon of the Star Point Sandstone.

- About 30 km further east, the Blackhawk Formation has almost entirely disappeared, being represented by only about 15 m of sandstone at the base of the Castlegate. The great thickening of the Mancos Shale outcrop has also altered the dominant footslope processes from gullying and pedimentation to massive slumping, at least on the upper section of the outcrop. In Nash Canyon, the sandstone cliff has been incorporated in extremely large rotational failures seated in the Mancos. At Westwater Canyon, not only has the Blackhawk Formation pinched out, but the Castlegate Sandstone no longer forms the upper cliffs of the escarpment. Instead it crops out as a low cliff and narrow bench deep within the Mancos Shale.
- Still further east, at Mount Garfield, the Castlegate Sandstone has pinched out, and the cliffs capping the Mancos Shale are cut from the Sego Sandstone, which lies about 150 m above the horizon of the Castlegate. In general, the shaley to sandy Nelson and Farrer Formations, which become increasingly prominent as the Bluecastle sandstones pinch out, are stripped back from the underlying Sego Sandstone.

However, in places like Crescent Canyon, where the shales are bolstered by a local thickening of sandstone in the Farrer Formation, the Sego Sandstone is capped by cliffs and benches developed in these stratigraphically higher units. Topographically the Mancos Shale expands from merely forming footslopes, to progressively occupying more and more of the escarpment face. These lithological changes are also reflected in the planimetric form of the escarpment. East of the Green River, where the shales dominate, the Book Cliffs have an irregular front extensively dissected by canyons; west of that river, where sandstones are more prominent, the line of the cliffs is much more regular. The Cretaceous sandstones of Utah also illustrate just how rapid facies changes may be within a single formation, especially near the source of sediment (Spicker, 1949). At its type locality, on the eastern edge of the Wasatch Plateau, the Castlegate Sandstone consists of about 150 m of massive cliff-forming sandstone; but about 15 km to the west, its coarser texture and more irregular bedding is reflected in the irregular and vertical faces broken by minor benches. Thereafter it changes rapidly to a very coarse conglomerate. The North Horn Formation of the adjacent Gunnison Plateau changes even more rapidly than the Castlegate. In the centre of the plateau, this formation is composed of about 760 m of variegated shales with subordinate sandstones. But a little more than 1 km to the south, the entire 760 m consists of a hard sandstone with many conglomeratic beds. A few kilometres further south, the entire formation is different again, consisting of shales and siltstones with some limestones. A little to the northwest, it consists of red beds with no sign of the massive, cliff-forming sandstones. As Spicker emphasized, it is only the excellent and virtually continuous exposures of this formation that allow these abrupt changes in facies to be deciphered, yet their topographic importance is very great.

Murchison River, Western Australia

Outcrops of the Tumblagooda Sandstone in the lower Murchison River valley of Western Australia illustrate yet another geomorphological aspect of facies change (Young, 1983a). The varied facies of this quartz arenite reflect shifts in the boundary between fluvial and marine depositional environments that have been mapped in detail by Hocking (1980). Hocking recognized three major facies associations. The first is mainly a trough cross-stratified, medium to coarse sandstone of mainly fluvial origin, which has fining-upward cycles 2–10 m thick. The second is a thinly bedded, fine to medium sandstone of marine origin, with generally planar bedding. The third consists of fining-upward cycles 10–15 m

Fig. 2.19 Cliffed upper slopes and thinly benched lower slopes on the Tumblagooda Sandstone, Murchison River, Western Australia

thick of dominantly coarse-grained sandstones, which, near the modern coastline, grade upwards into red siltstones.

- In the upper reaches of the Murchison gorge, cliffs are cut only in the first of these facies associations. Failure there is mainly by collapse of massive jointed-bounded blocks, especially where they are undercut by the river or by the enlargement of seepage cavities on major bedding planes. There is also evidence of block sliding down inclined bedding planes, although the maximum dips, where the sandstone is locally tilted, are only 22°.
- In the central part of the gorge, where the first facies association is underlain by the second, there are prominent cliffs on the upper slopes but numerous small benches and risers on the lower slopes (Fig. 2.19). As the first association pinches out downstream, the slopes are increasingly dominated by small benches and risers, with large vertical faces occurring only where outcrops undercut by the river have failed along major fractures. These vertical faces then degrade progressively into stepped forms that are lithologically controlled by the upward-fining cycles.
- In the third facies association, especially where the fining cycles terminating in siltstones dominate, two types of slope assemblage can be seen. Where the siltstones are thickest, cliffs a few metres high are separated by broad benches that have widened along the outcrop of the siltstones. But where the siltstones are thin, the slopes again consist of numerous small benches and risers. In some places, where the siltstones are very thin, the benches are covered with irregular hummocks of slabby sandstone that have accumulated as debris collapsed from the adjacent riser.

Conclusions

The characteristics of sandstones, and thence the landforms developed on them, vary in response to a range of factors, including the original sedimentation, changes during lithification and tectonic history. These variations are important not only between sedimentary basins, but both vertically and laterally within them. Failure to recognize the effects of variations such as facies changes can lead to erroneous geomorphological interpretation. When attempting to correlate erosional surfaces in southeastern Australia with the sequence of surfaces that he had proposed for Africa, King (1959) mapped a series of mesas west of Yalwal, in the Shoalhaven catchment, south of Sydney, as remnants of an Early Tertiary pediplain. The absence of similar mesas further east in the catchment was supposedly due to deep incision of the Shoalhaven River and its tributaries during a subsequent cycle of erosion triggered by uplift. Their preservation west of Yalwal is, however, not due to variations in erosion cycles across the catchment but to a facies change in the Permian Berry Formation that constrained the pattern of denudation (R.W. Young, 1977a). In the eastern part of the catchment, the Berry

Formation consists mainly of siltstones and shales deposited in relatively deep water, but as the former Permian shoreline is approached, the formation becomes increasingly sandy. It is in the sandy, much more resistant facies that the mesas are cut; the softer siltstones nearer the modern mouth of the catchment have been deeply dissected. In short, the mesas are structural-lithological features, not remnants of a pediplain.

In the following two chapters, we consider the landform elements of cliffed and curved slopes, discussing the processes that shape them.

3

Cliffs

Northwards from the rim of the Grand Canyon of the Colorado River, a great stairway of cliffs and benches rises more than 1500 m to the summit of the high plateaux of southern Utah. Most of the major cliffs on this stairway are cut in sandstones. Other huge series of cliffs cut in sandstones are to be found in such diverse settings as the Hombori of the southern Sahara and the Roraima of the Guyana shield. Smaller, but still visually very impressive sandstone cliffs occur in places like the Kimberley Plateau of northwestern Australia, the Sydney Basin of the southeastern part of the same continent, the margins of the Drakensberg of southern Africa, the Torridonian Mountains of western Scotland, and the sandstone lands of central Europe. In short, bare rock slopes, many of which stand in vertical faces, are major features of most sandstone landscapes and are the dominant feature of many of them (Fig. 3.1).What, then, are the rock characteristics and forces that shape these cliffs?

The most basic control on the height and length of cliff is the thickness and extent of the sandstone strata. Universally, the force of gravity works to destabilize the cliffs as the weight of rock tends to crush the basal layers, or masses of rock fail and fall from the face. Tectonic forces can operate slowly – for example, as locked-in stresses are released in faces exposed by erosion; or they may act rapidly – for example, during earthquakes. Climate, especially via its role in determining the flow of water across and through cliffs, is an important factor. And within the clifflines, variations in rock type, folding, jointing and fracturing, bedding planes and other sedimentary or diagenetic or weathering features can influence the detailed shape and extent of the cliffs.

Strength–stress relationships in cliffs

In his classic study of rock slopes, Terzaghi (1962) noted that cliffs in hard rock never reach the heights that simple theory might predict. If values for the

51

Fig. 3.1 Cliff face produced by the collapse of joint-bounded blocks in the Blue Mts, western Sydney Basin. Although the rock face appears fresh, the collapse occurred 70 years ago

compressive strength (*S*) and unit weight (*W*) typical of sandstone are substituted in Terzaghi's simple equation:

$$H = S/W \qquad\qquad (3.1)$$

theoretically the critical height (*H*) which a sandstone cliff could attain before collapsing under its own weight is about 1700 m. No vertical sandstone cliffs of that height exist! And indeed, no cliff in any rock is so high – the world's highest near-vertical cliffs are those in granite of Trango Tower (1340 m) in the Karakoram, and the highest vertical cliffs (again in granite) are on Mt Thor on Baffin Island (1250 m). Of course, as most sandstone outcrops are much thinner than 1700 m, cliffs on them cannot reach the critical height at which the rock would crumble under its own weight. Sandstone can therefore be expected to stand in steep faces. But this is not the whole story. As Terzaghi emphasized, the strength of rock masses depends not only on the strength of the rock, but also on the fractures and discontinuities through it.

Hoek and Brown (1997), reviewing their empirical scheme for estimating rock mass-strength, noted that geotechnical software is still usually written in terms of the Mohr–Coulomb failure criteria relating cohesive strength and angle of friction (shearing resistance) as determined from triaxial testing. These criteria clearly relate to the rock sampled. The strength of the rock mass, however, depends also on other factors, such as strength reduction due to porewater pressure or to dehydration causing deterioration of cement, and the freedom of blocks of intact rock to move past one another along joint or bedding or fold planes.

When an unfractured mass of rock is compressed, if the force is great enough, it may be crushed, or it may shear across the block. In sandstone, this shearing happens before the force is great enough to crush the rock. The shearing strength of sandstone is generally less than about half the uniaxial compressive strength. Thus failures on natural rock faces have frequently been explained in terms of simple models of shearing failure developed as a result of laboratory experimental work. Such assumptions may be valid, but caution is needed. Although normally much lower than compressive strengths, shearing strengths of intact sandstone are generally sufficient to withstand stresses generated in most cliffs, unless, of course, the rock fails along pre-existing fractures. Certainly it should not be assumed that natural rock outcrops will initially fail along planes similar to those generated in triaxial testing. The high confining pressures of triaxial testing ensure a shearing failure at about 45° to the direction of the primary stress. On theoretical grounds alone, natural outcrops should not be expected to do so, for the face of the cliff is not confined but acts as a free boundary.

Brittle fracture

Rocks are much less resistant to tensional stresses than they are to compressive and shearing stresses. The ratio of compressive to tensional strength in sandstones is generally in the range of 10–16, averaging 15 (McNally and McQueen, 2000). However, the ratio in soft sandstones is only about 8–10. It may be that shearing, rather than true tensile failure, occurs in these non-brittle rocks. On the other hand, high ratios (16–20) occur in much harder rocks such as quartzite; but because these rocks are brittle, once fractures are initiated, they propagate rapidly (McNally and McQueen, 2000). Thus, brittle fracture across blocks could be expected in any type of sandstone where tensile stresses are high, and there are several ways in which this can occur.

The simplest case is where part of the face is undercut, but is still joined to the main mass. The undercut section will fail when the gravitationally induced stress exceeds the tensile strength. If we assume that the block drops vertically, theoretical estimates can be derived readily for the critical height of the undercut mass by inserting values for tensile, rather than compressive strength in Terzaghi's equation. Robinson (1970) estimated that for the Navajo Sandstone, with a density (unit weight) of about $2300\,kg/m^3$ and a tensile strength of 1.2–3 MPa, a column with a cross-section of about $6\,m^2$ could be supported in tension to a theoretical height of about 70 m. Judging from the size of the recessed scars on cliff faces, he suggested that this was a realistic calculation, although he also warned that fractures in the column could cause it to fail before the critical limit. This type of calculation could be repeated for sandstones of varying strengths, but the magnitude of the tensional stress clearly depends on the assumed sectional form as well as the thickness of outcrop. For example, a very slender undercut column of a typical Sydney Basin quartz sandstone, with a density of $2500\,kg/m^3$ and tensile strength of about 5 MPa, could be supported to the full height of a 150-m cliff, whereas a thicker column would break and drop from the cliff face.

Where the undercut section of sandstone takes the form of a block or plate projecting from the cliff, rather than a column, its stability can be more appropriately analysed by considering it to be analogous to a simple cantilever. Tensional stresses at the junction of the projecting plate and the cliff will again be determined by the density of the sandstone and the dimensions of the plate (Fig. 3.2). The stresses arise from the load $P = 2cl\rho$, where ρ is density, acting along a moment arm of length $l/2$ of a cantilever of thickness $2c$. Following the method of Timoshenko and Goodyer, Robinson (1970) derived the tensile stress acting along the junction of the plate and the cliff, and then rearranged the equation to give a ratio of the plate dimensions at the critical level at which failure occurs:

$$l/c = (2\sigma/3\rho c)^{1/2} \tag{3.2}$$

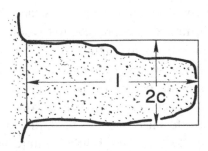

Fig. 3.2 The parameters used in Equation (3.2), representing a projecting plate as a cantilever

Fig. 3.3 Eagle Rock, Royal National Park near Sydney, Australia. The projecting plates are locked in by wedge-shaped bedding planes, and are longer than most overhangs in the Hawkesbury Sandstone

where σ is the tensile strength. Assuming a density of 2.3 g/cm^3 (= 2300 kg/cm^3), a tensile strength of 2 MPa, and letting $2c = 40$ m, Robinson computed a critical length of 34 m for projecting plates in the Navajo Sandstone. Estimates by us for Sydney Basin sandstones, with a tensile strength of 5 MPa and projecting in a plate with $2c = 10$ m, give a critical length of about 25 m. Again, however, the theoretical limits normally exceed observed examples, for projecting overhangs in the Sydney region (Fig. 3.3) are rarely more than 10 m, unless they are supported at both ends (for which alternate forms of analysis must be used). As was the case with Terzaghi's computations of critical height based on compressive strength, the dimensions of undercut outcrops depend on more than just the strength of the rock. Either there is an increase in the local magnitude of the

tensile stress, or the rock fails primarily along micro-fractures. These compli-
cations are dealt with later. The calculations presented here serve primarily as
an expression of the general nature of the relationships between the forms and
materials of cliffs, and must now be modified to account for the actual com-
plexities encountered on sandstone faces.

Horizontal (lateral) stresses

Tensile stresses in a rock face are by no means limited to sections that are
undercut; they may operate over large sections of a slope, and must therefore be
considered as part of a general boundary effect. The generation of such tensile
states can be considered first for a simple valley side. Every type of solid changes
shape by stretching or contracting as force is applied to it. Not only does it
expand or contract along the primary axis, it also expands or contracts sideways.
If we now return to Terzaghi's simple demonstration, we see that the weight of
the rock not only causes vertical compression, but also causes secondary,
or lateral deformation by expansion. As a block of rock is compressed, it also
bulges – imperceptibly to the naked eye – near the base. The relationship between
the strain, or deformation, in the direction of the primary axis and the strain at 90°
to that axis is given by Poisson's Ratio. The range of values for the Poisson Ratio
encountered in sandstones is from about 0.2 to 0.4 (Lama and Vukuturi, 1978).
The lateral boundary force set up in this way is given by the equation:

$$G = \frac{v}{1-v} \rho g H \tag{3.3}$$

where G is the gravity-induced lateral stress, v is Poisson's Ratio, ρ is density and
H is the height of the valley side. For example, the lateral stress generated at the
base of a 100-m cliff of sandstone, with a density of $2500 \, \text{kg/m}^2$ and a Poisson
Ratio of 0.4, is approximately 1.63 MPa, or about 25–50% of the tensile strengths
for these rocks (Jaggar, 1978a; Pells, 1977).

   The lateral stress induced simply by the weight of the rock may be increased by
tectonically induced stresses acting approximately in the horizontal plane. For
example, measured stresses in an underground coal mine, south of Sydney,
Australia, could be resolved into a vertical or gravitational stress of 12 MPa and
horizontal or tectonic stresses acting at 90° to one another with magnitudes of 25
and 35 MPa (Jaggar, 1978a). As these measurements were made only about
200 m below the surface, the measured horizontal stresses must also act on the
cliffs of gorges cut in the sandstone plateau under which the mine extends.
Studies of stress fields in the Sydney area (Braybrooke, 1990; McQueen, 2000)
show that from depths of 0–20 m, major horizontal stress consists of the gravity

component of overburden and a tectonic component of 2.5 MPa. From 20–200 m depth, the tectonic component increases to 6.5 MPa. The tectonic component of the minor horizontal stress for these range of depths is 2 MPa and 4.5 MPa, respectively. The orientation of the major horizontal stress is mainly to the northeast and generally does not change systematically with depth.

However, topography does exert localized influence on the regional trends. Below some palaeochannels in the Hawkesbury Sandstone, major horizontal principal stresses are 6–9 times greater than the vertical stress; while in the Blue Mountains west of Sydney, the orientation of the major horizontal principal stress is locally parallel to a 200-m high east–west scarp (McQueen, 2000). Excavation in sandstones and shales for the Mangrove Creek Dam, north of Sydney, revealed a 'false anticline' of bulged rock with flanks dipping at 4–5°, and with a steepening of the dip in the core of the structure to 45° (McNally, 1981). As load had been removed by valley cutting at this site, the rock had bulged upwards, a conclusion supported by the occurrence of shear zones sub-parallel to the bedding and extending to 25 m depth below the valley. In the gorge in the Blue Mountains across which Warragamba Dam has been built, there has been stress release on the gorge wall, but increased stress below the valley floor (McQueen, 2000). While this occurs naturally due to erosion, it can be accentuated by the subsidence impacts of underground mining (Fig. 3.4).

Fig. 3.4 Thin blocks of sandstone pushed up into a tent-like shape by compressional stresses due to subsidence over an underground coal mine, southern Sydney Basin

Measurements in tunnels through sandstone in the Sydney region show that distribution also varies with the stiffness or elasticity of the rock (McQueen, 2000). At one site, the major horizontal stress was 8.1 MPa with an orientation 120° at a depth of 60 m in massive sandstone, but was 5.2 MPa at 50° in sandstone of lower stiffness at 50 m. In another tunnel, stress was redistributed from low modulus siltstone to stiffer sandstone. Failure occurred due to subsequent shear slip of beds and compressive failure in thin incompetent layers of sandstone.

The greater the tectonic component of stresses on a slope, and the steeper the slope, the more extensive is the zone of tensile stresses behind the slope. Extensive tensile zones will develop wherever lateral stresses are 3 times greater than vertical stresses and the slope is steeper than 45° (Stacey, 1970). These conditions are likely to be fulfilled along many sandstone cliffs. At the foot of cliffs, or other abrupt breaks of slope, undercuts or indentations, stress is concentrated. Gerber and Scheidegger (1973) argued that rock walls break down at the base for that reason, but this is overly simplistic, as the tensile zones develop well above the base. Stress trajectory diagrams allow visual estimation of the pattern of stress concentration around indentations or undercuts on cliffs; the pattern varies with the shape of the indentation, and with the ratio of vertical to horizontal stresses (Hoek and Brown, 1980).

Undercutting

Given the importance of undercutting in creating concentrations of stress in sandstone cliffs, its origins need to be understood. In many instances, the cause is direct erosion by stream or wave action and requires no further comment here, but, in the majority of instances, the cause must be sought in other slope processes. The standard explanation attributes undercutting to the concentration of seepage at the contact between permeable sandstones and underlying and less permeable claystones or shales. The seepage can have several obvious effects. Firstly, it increases local porewater pressure which may contribute to failure. Secondly, it can produce swelling of clays and plastic failure in the fine-grained sediments. Thirdly, the flow of groundwater may leach the cement between the sand grains, producing an irreversible decline in the strength of the rock. There are many instances of these kinds of changes caused by seepage in undercut sections of cliffs, but in our experience, most undercutting is not caused by these processes.

The majority of recesses developed along shale or claystone beds that we have seen below sandstone cliffs involve brittle fracture, not plastic deformation (Fig. 3.5). Unless seepage is concentrated along major joints, the really intense weathering of the clayey rocks is located not in the recess, but beyond the drip line of water coming over the cliff, where it promotes mass failure of saturated

Fig. 3.5 Claystones under a sandstone overhang failing by brittle fracture. Note that the claystone is barely weathered

clays downslope from the recess. The claystones in the recess are certainly altered, by seepage or subaerial weathering, but the fragments are generally still quite resistant to plastic deformation. Only minor hydration of clay seems to have been needed for these rocks to fracture. The detailed review given by Yatsu (1988) leaves no doubt that pressures generated by the swelling of clays can be very substantial, especially for montmorillonites. Furthermore, even intact rocks may lose much of their strength when wet. For example, Schmidt–Hammer readings on shales in a cliff-foot recess immediately below a Hawkesbury Sandstone cliff on the Illawarra Escarpment, south of Sydney, showed a 45% reduction in strength between a dry outcrop and an adjacent wet outcrop of the same intact shale. The generation of stress and the reduction of strength due to the wetting of shales and claystones seems to be the cause of the brittle fracturing of these clay-rich rocks.

Nonetheless, close inspection of many undercut faces reveals that it is not always the clayey rocks, but rather the sandstones immediately above them that are the focus of disintegration and the formation of a recess. The primary cause again seems to be a reduction in the strength of the rock. Pells (1977) has demonstrated that the uniaxial strength of the Hawkesbury Sandstone is reduced by more than 50% from a dry to a fully saturated condition. Studying English sandstones, Priest and Selvakumar (1982) reported a drop in strength for the

Bunter Sandstone from 57 MPa to 38 MPa for only 1% increase in moisture content above a totally dry state; and Dobereiner and de Freitas (1986) note that the strength of saturated Kidderminster Sandstone is only about 20% of the dry strength. Brighenti (1979) observed that the strength of wet Tuscan Apennine sandstones was about 70% of the dry strength when in compression, but only about 50% when in tension. Undercutting is often concentrated where conglomeratic or brecciated beds occur at the base of a sandstone or where these beds are inter-bedded with the sandstone. The preferential breakdown of very coarse beds may well be due to their higher permeability, which concentrates the seepage and promotes the weakening of any cement in the matrix between the clasts, causing the clasts to drop out.

Undercutting may be both partly triggered and also extended as a result of stress concentration at the cliff base. The lateral stresses at the cliff base may be similar to the strengths of the rocks there, especially if these are relatively weak rocks like claystones. In this situation, little weakening by weathering or seepage is needed to trigger undercutting. Furthermore, it is likely that the role of brittle fracture will increase as the recess deepens because, as the weight of the overhanging mass is carried on smaller and smaller sections, stresses increase progressively. This elementary, though neglected point, can be readily illustrated. If a cliff is transmitting a stress of 2.5 MPa to the underlying claystone at the beginning of undercutting, the average stress level – with no additional allowances made for local stress concentration – will have risen to 5 MPa when 50% of the section is undercut; and it will continue to rise in a geometrical progression until the supporting rock fails, and the overhanging mass collapses.

Of course, the opposite effect can be expected if, instead of being undercut, the cliff is buttressed outwards at the base. Buttressing will result in a reduction of the footslope concentration of the gravitationally generated part of the load. When the vertical load is combined with any lateral tectonic push, buttressing of a cliff can be seen to be roughly analogous to the standard buttressed forms of dams, or to the flying buttresses of cathedrals. Here we need consider not just the distribution of stress in relation to the strength of the material, but also the position of the centre of gravity, or the thrust line, of the load. In short, the thrust line of the load tends to be deflected outwards by the horizontal component of the load; but, as long as it does not extend beyond the junction of the vertical face and the valley floor, the cliff will be stable.

Fatigue and weathering effects

Cliffs are subjected to stresses not just once but many times. Over the time scale of landform change, repeated cycles of stresses are applied at or near the surface of outcrops. The accumulated effect of these cycles may eventually cause the rock to fail at a stress level – the fatigue limit – well below the strength determined by conventional testing. Weakening of sandstone by fatigue effects was demonstrated by Burdine (1963), in his experiments on the cyclical compressive loading of Berea Sandstone. He found that the difference between the fatigue limit and the static compressive strength of this sandstone increased as the confining pressure fell. At very low confining pressures (0.2 MPa) the fatigue limit was 24% below the static strength, a reduction that was matched by the 20% difference between fatigue and static strengths of the Tennessee Sandstone obtained by Hardy and Chugh (1971). Burdine also showed that the compressive fatigue limits for wet samples were consistently lower than those of dry samples. Brighenti (1979) demonstrated similar differences between wet and dry samples of Tuscan sandstone, not only in compression but also in tension; the fatigue limit for wet samples in tension was only about 40% of the static tensional strength. Moreover, Haimson (1974) showed for a variety of rocks, including the Berea Sandstone, that the fatigue limit of dry samples in cycles of tension–compression was considerably lower than in tension alone, falling to about 30% of the static strength. Not only seepage but also microclimate needs considering, because experimental work by Winkler (1994) demonstrated a significant reduction in the strength of sandstone following exposure to relative humidity of only 15% and above. These results provide a useful guide to the likely changes in the strength of sandstone outcrops.

Repeated wetting and drying, and consequent expansion and contraction of clays in the matrix, is a significant cause of fatigue in many sandstones. Changing capillary tension (negative porewater pressure) is the main reason why clays harden on drying and soften on wetting. A film of water drawn into the finest pores acts as a confining membrane that exerts maximum capillary pressures in the order of 100 MPa; but saturation reduces this pressure to zero. As clay also reduces permeability, it may inhibit the development of intergranular bonding. These effects vary with the percentage and type of clay present. The effect is relatively slight in sandstones in which kaolinite is the dominant clay, as there is limited shrink–swell in response to changing water content in kaolinitic clays. Montmorillonitic clays are more likely to cause loss of rock strength. The presence of swelling clays can play a major role in the disintegration of laminated sandstone, for those laminae containing much smectite break down more readily than those laminae containing little clay. An extreme case was reported by Teruta (1963), who demonstrated that arkose at the site of the Prek Thot Dam in Cambodia is unstable because of the remarkable swelling of a matrix dominated by montmorillonite and mixed-layer clays. However, where kaolinite comprises a high percentage of the total volume of the rock (up to 40% in some beds of the

Hawkesbury Sandstone), it may be responsible for substantial creep of some sandstones under sustained pressure (Johnson, 1960; Pells, 1977).

Also, it has become increasingly clear that brittle fracture of rock is often influenced greatly by chemical weathering. At the macroscale, the measured strengths of sandstones are generally less for weathered than for unweathered specimens. This is especially true where calcite is the dominant cement. In some instances, however, moderate weathering may lead to an increase in strength. Johnson (1960) noted that in the Hawkesbury Sandstone near Sydney, fine-grained, sideritic or shaley beds have their strength reduced by weathering, whereas coarse-grained, quartzose beds often actually become stronger when moderately weathered. He attributed the strengthening of the second group to the forming of additional limonitic cement by the oxidation of siderite.

Mechanisms of brittle fracture

Brittle failure in rocks has most commonly been explained in terms of Griffith's theory of crack propagation, which emphasizes the concentration of stress along microscopic cracks and the consequent expansion of those cracks to a critical size that triggers catastrophic failure (Griffith, 1921). This approach is summarized very briefly here.

The numerous cracks in materials will remain stable provided that they do not reach a critical length; thereafter they tend to propagate and cause catastrophic failure. The critical length (L_c) can be expressed simply as:

$$L_c = \frac{1}{\pi} \cdot \frac{\text{Work of fracture per unit area of crack surface}}{\text{Strain energy stored per unit area of material}} \qquad (3.4)$$

or $\qquad L_c = \dfrac{2\omega E}{\pi \sigma^2}$

where ω is the work of fracture for each surface, E is Young's Modulus and σ is the average tensile stress in the material near the crack. The work of fracture is the energy required to break a given cross-section of material. This requirement is supplied by the energy stored in a material as it is strained. The concept of strain energy is best visualized by the energy stored in a wound spring or a drawn bow. For a crack to propagate, the available strain energy must equal the work required to tear apart the bonds which hold together the material at the tip of the crack.

To what extent the theory of crack propagation can be applied to the failure of rocks is still debated. Hoek (1968, 1983) and Hoek and Brown (1980) developed a general failure criterion, derived from a modified form of the Griffith criterion, that fits experimental data very well indeed; and the importance of crack propagation to the understanding of geomorphological phenomena in igneous rocks was restated by Whalley *et al.* (1982). Nonetheless, strong reservations about its applicability to failure in rocks have been expressed (e.g. Rudnicki, 1980). Experimental studies by Lajtai (1971) indicate that the initial mode of failure around cracks of any shape and orientation is the appearance of tensile fractures originating at the point of maximum tension and propagated approximately parallel to the applied uniaxial load; the shearing fractures predicted by the Griffith theory develop later. Jaeger (1971, p. 103) summed up the situation succinctly by noting that:

Griffith theory has proved extraordinarily useful as a mathematical model for studying the effects of cracks on rocks, but it is essentially only a mathematical model; on the microscopic scale rocks consist of an aggregate of anisotropic crystals of different mechanical properties and it is these and their grain boundaries which determine the microscopic behaviour.

This is especially true of porous agglomerations of granular material such as sandstones. Our experience with scanning electron microscopic (SEM) observation of surfaces of rupture in sandstones supports Jaeger's general observation; although fractures are certainly propagated across grains, many develop through the sutures or cement bonding grains. To this must be added the fact that weak sandstones fail not by the fracture of grains or cement, but by the rotation and displacement of grains (Dobereiner and de Freitas, 1986).

Notwithstanding the uncertainty which still surrounds the precise mechanism of failure, there are similarities between macroscopic failure across blocks of sandstone and the surface patterns produced by the propagation of fractures that have been well documented at the microscopic level in studies of ceramics. The typical pattern of failure in tension that develops at the microscopic level is a semi-circular array of surface textural variations around the initial flaw.

- First, there is a zone in which the flaw propagated to the critical size.
- Then follows the fracture 'mirror', a flat, smooth zone formed in an accelerating phase of crack growth after the tensile strength is exceeded. In weak materials the mirror may extend across the entire remainder of the fracture surface, but in strong materials it gives way to distinct zones of 'mist' and 'hackle'.
- The mist is a rough or stippled zone, and the hackle is a very rough area of lines that are radially aligned to the critical flaw at the origin of the fracture.

The geomorphological significance of the surface textural patterns is that they allow us to reconstruct the manner in which the fractures have propagated.

In many instances, arcuate patterns extend across the entire failure, with the focus of the arcs at the top of the outcrop. This pattern indicates that the failure was propagated downwards from a tension crack which developed on the upper

Fig. 3.6 A curved scar in Aztec Sandstone, The Valley of Fire, Nevada. The alignment of the curvature indicates that the fracture propagated laterally from the joint on the right-hand side of the outcrop

slopes, rather than from a fracture extending upwards from a notch at the base. Such failures can be accounted for by the frequently cited model developed by Hoek and Bray (1974) for tensional failure triggered by the undercutting of a rock face. In this model, a previous failure is assumed to have left a tension crack behind that face at the top of the slope. As a result of renewed under-cutting, a new failure plane develops downwards to the undercut section from the base of the tension crack.

Not all failures fit this model. Figure 3.6, photographed in The Valley of Fire National Park, Nevada, shows a block of Aztec Sandstone which has failed in tension from a face bounded by a vertical joint. The surface of failure is marked by very distinct arcuate lineations and hackle striations that extend radially from approximately the mid-point of the adjacent joint plane. This pattern shows that the failure propagated not downwards from a tension crack on the upper surface as predicted by the standard model, but laterally across the face. Moreover, as the arcs and radial striations cut across the prominent horizontal bedding of the sandstone, this failure was clearly a brittle cleavage that was independent of textural variations in the rock.

These clear contrasts in the direction of fracture propagation prompt an assessment of the extent to which Hoek and Bray's model of slope failure in tension can be applied to the interpretation of sandstone landforms. We have analysed the direction of propagation, as indicated by surface markings, in a set of 40 brittle fractures in sandstones from a range of locations. All of the cases considered are due to the brittle fracture of intact rock on canyon walls, and are not simply faces from which joint-bounded blocks have fallen. The data are arranged in four groups – those in which the direction of propagation is dominantly upwards (oriented between 316° to 45° in the vertical plane), dominantly right lateral (46° to 135°), dominantly downwards (136° to 225°) and dominantly left lateral (226° to 315°). Although the data set is small, and needs to be duplicated by further studies, the results show clear trends in the apparent direction of fracture propagation. 50% of fractures extended upwards, 25% were right lateral, 15% left lateral, but only 10% were oriented downwards. The results seem to indicate that the failures were triggered by local concentrations of stress that reflected variations in the morphology of cliffs, rather than by any general tensile stress field such as that assumed in the standard model of failure triggered by undercutting.

Block failure

Instead of failing by brittle fracture through an undercut section, an entire joint-bounded block may be displaced and fall. That is to say, failure may not be from lack of strength but from lack of stability. The simplest instance is that of horizontally bedded and vertically jointed rock, which behaves in a manner similar to a stone wall, and thus can be analysed in terms of Thomas Young's classic study of masonry. Young (see Gordon, 1978) pointed out that when a block is undercut, the load carried by a block becomes increasingly asymmetrically distributed. Until the centre of the load reaches the edge of the 'middle third' of the block, the stresses across the face of the block remain compressive. Thereafter, tensional stresses will be set up on the inner face of the block, and the block will begin to tilt outwards. Once the centre of the load acts beyond the outer edge of the wall, the block will hinge, tip up and topple. This emphasis on the critical 'middle third' assumes that the block is free to hinge and is not held in place by adjacent blocks, and that its base is horizontal. This is not always the case, as joint blocks on a cliff face may be wedged in by adjacent blocks, especially if the jointing is not exactly rectilinear.

Whether or not the centre of the load falls beyond the confines of a block, causing the block to topple, depends on the geometry of the block, and specifically on the ratio of its width (w) to height (h), and on the angle (Ω) at which it is inclined (Goodman and Bray, 1976). Toppling will occur when:

$$w/h < \tan \Omega \qquad (3.5)$$

The inverse form:

$$w/h > \cot \Omega \qquad (3.6)$$

is also widely used (Wylie, 1980).

The simple trigonometrical relationships expressed in the limiting equation for toppling show clearly why high and very slender towers carved from sandstones remain stable unless undercut at the base (Wylie, 1980). The Old Man of Hoy, a 137-m high sea stack cut from the Old Red Sandstone of the Orkneys, is a case in point. Although its height exceeds its width by a factor of about 6 or 7, this stack would not topple until inclined to about 10° from the vertical. As the dip of the sandstone is almost horizontal, substantial under-cutting must occur before the stack will collapse. The great towers of the De Chelly Sandstone of Monument Valley are similar instances. Despite their very slender form, with the height of some of them exceeding their width by a factor of 10 or more, the dip of the beds is less than the critical declivity of between 5° and 6° at which they would topple.

Sandstone blocks may also topple backwards if basal failure occurs. For example, closely spaced jointing in the sandstones of the Mesaverde Group of the southwestern USA has produced columns that have dropped vertically, but have also tilted backwards against canyon walls (Schumm and Chorley, 1966, photo 19). In eastern Australia, Cunningham (1988) pointed out that subsidence due to mining under cliffs at North Nattai coal mine, southwest of Sydney, caused sandstone towers up to 90 m high to move downslope and to rotate backwards. The broken blocks on the talus slope below the cliffs are still arranged according to the stratigraphy, with blocks from the basal strata furthest downslope and those from the upper strata resting close to or on the base of the cliffs. Using Goodman and Bray's analysis, Cunningham calculated that the towers would have been tilted to about 13° before they toppled backwards. Reverse toppling seems to be triggered either by a curved basal failure plane, or by a block geometry which causes the centre of the displaced mass to lie to the rear of the vertical axis (Fig. 3.7).

The simple trigonometrical causes of toppling show that it is likely to occur where jointing, rather than bedding, dominates the fracture pattern. Where fracturing along bedding planes is dominant, block sliding or erosional quarrying is more likely than toppling. Changes in the geometry of bedding and jointing may thus lead to significant variations in the mode of failure on an individual

Fig. 3.7 A back-tilted block of Nowra Sandstone, Yalwal, southern
Sydney Basin

cliffline. The sea cliff near the Brough of Dearness in the Orkney Islands is a case
in point (Fig. 3.8). The upper section of this cliff is dominated by joint-bounded
blocks elongated on the vertical axis, while the lower section is dominated by
bedding planes that produce blocks elongated on the horizontal axis. The average
width–height ratio of blocks in the upper section is about 0.2, and 2–3 for blocks
in the lower section. Blocks in the upper section are inclined close to their critical
toppling limit of about 12°; but at this inclination, beds in the lower section are
quite stable and could only topple if undercut significantly. The change in
geometry, and thence in the dominant mode of failure, is marked in the change of
slope form – the vertical rock face breaks sharply at its base to a series of benches
that dip seawards.

Fig. 3.8 A coastal cliff in Old Red Sandstone, Brough of Dearness, Orkney Islands. Vertical jointing is dominant in the upper part, producing tall narrow blocks that topple forward. Dominance of bedding in the lower part produces long flat blocks that must be undercut to the central third before becoming unstable

Block sliding

Where fracture planes are inclined downslope, sliding of blocks may contribute to the failure along clifflines. The relationship between the driving and resisting forces at which sliding is initiated is, in its simplest form, given by the standard Mohr–Coulomb formula:

$$\tau = c + \sigma \tan \phi \tag{3.7}$$

where τ is the shearing strength (i.e. the stress at which shearing is initiated), c is cohesion across the sliding plane, σ is the stress acting normal to the sliding plane (gravity), and ϕ is the angle of friction on the sliding plane (see, for example, Hoek and Bray, 1974). The terms are more fully defined, and the effect of porewater pressure taken into account, in the form:

$$W \sin \Omega + V = c + (W \cos \Omega - U) \tan \phi \tag{3.8}$$

where, in addition to the terms defined in the previous equation, Ω is the slope, W is the weight of the block, V is porewater pressure in the tension crack behind the block and acting in the downslope direction, and U is porewater pressure beneath the block.

The use of the effective stress, which takes account of water pressure, is generally necessary in sandstones, not only because of their role as major aquifers, but also because often the permeability of individual beds in a sequence of sandstones varies considerably. This means that there may be semi-confined aquifers within a sandstone sequence, and resultant variations in the stresses on the beds. Although the water pressures acting on bedrock slopes are generally quite small, the large areas over which they act can produce very substantial forces. For example, where upland swamps on a sandstone plateau direct subsurface flow towards a cliff face, the cliffs become degraded and broken; whereas in better-drained sections, the cliffs stand in high vertical faces (Young and Young, 1988). This is not due to undercutting, or basal sapping, but to sliding of blocks across one another right through the depth of the sandstone exposed on the cliffs.

Stress–strain plots for jointed rocks often display an initial threshold that must be exceeded before movement occurs. It is conventional to regard this threshold as akin to the cohesion factor in the Coulomb equation for the shearing resistance of soils, and in many cases joints in rock do have a true cohesion produced by partial cementing. Probably in most instances, however, the initial resistance is not a true cohesion due to cementing, but an apparent cohesion attributable to the surface roughness which must be overcome before sliding can commence. The cohesion intercepts for natural and artificially cut joints and bedding planes in sandstones are generally very low, ranging from zero to about 0.44 MPa (Jaeger, 1971; Pells, 1977; Jaggar, 1978b). Although still low, there is greater cohesion along contacts between sandstone and shale, than on sandstone-to-sandstone contacts; Jaggar (1978b) reports a cohesion of 0.9 MPa for a sandstone–shale contact in the Sydney region (Fig. 3.9).

The frictional resistance to sliding is determined essentially by the surface roughness, especially by the size, angle of inclination and the degree of interlocking of grain-sized projections, or asperities. At low to medium levels of applied stress, resistance occurs as the asperities slide over one another. As stress increases, the asperities begin to shear off and the rate of displacement rises until a new threshold is reached. Then displacement becomes virtually continuous with no additional increase in applied stress, presumably because of plastic deformation (Jaeger, 1971; Lama and Vukuturi, 1978). The angles of friction for the linear sections of the stress–displacement curves for sandstone-to-sandstone contacts range from about 27–44° (Jaggar, 1978b; Lama and Vukuturi, 1978).

Because of the shearing of asperities caused by any previous displacement, and also because of weathering of the rock surface, the frictional resistance on a sliding surface may be significantly lower than that on a fresh surface. Such residual values for joint faces (ϕ_{jr}) may involve a reduction of 30% or more. For

Fig. 3.9 A wedge of Hawkesbury Sandstone, defined by a channel base below and a horizontal bedding plane above, has slid out of an excavation

example, a sandstone with ϕ of 30° determined on a dry unweathered surface will have, if only slightly weathered, a ϕ_{jr} value of about 25° or, if severely weathered, a ϕ_{jr} value of about 20° (Barton and Choubey, 1977; Selby, 1982).

The effect of water on the frictional resistance of the sliding surface is variable. Water acts as an anti-lubricant on massive crystals such as quartz, but as a lubricant on layered-lattice crystals (Lama and Vukuturi, 1978). Therefore the effect on sandstone, which contains both massive and layered-lattice crystals, is complex. Jaeger's experimental work with dry and soaked specimens of sandstones showed a slight increase in the angle of sliding friction for the wet surface (Jaeger, 1971). Results different from those of Jaeger came from experimental work with the Hawkesbury Sandstone by Pells (1977). Pells found that, while there was no change in the frictional angle between the dry and wet states (both being 40°), the cohesion intercept fell from 0.44 MPa for a dry surface to zero for a saturated surface.

The stresses tending to shear asperities vary with the inclination of the surface of sliding in relation to the direction of the major principal stress (Jaeger and Cook, 1969). The stress required for failure will be at a maximum when the plane is inclined at close to 0° or 90° to the major stress, and a minimum when it is inclined at about 30°. Selby (1982) has approached this problem by pointing out that the depth of overburden required to initiate failure along a joint by shearing off asperities decreases with the inclination of the joint. Joints inclined at 75° will require an overburden in excess of about 300 m to generate the required stresses, while those inclined at 35° may require only 10–50 m of overburden.

The analysis of frictional resistance is more complex where joints intersect and a block between them fails as a wedge sliding along the two surfaces. In the simplest instance, on a fully drained slope and with zero cohesion, the factor of safety (F) for a wedge failure can be estimated by the equation:

$$F = A \tan \phi_A + B \tan \phi_B \qquad (3.9)$$

where A and B are empirical coefficients varying with the angle and direction of dip of the two intersecting faces, and ϕ_A and ϕ_B are the friction angles on the two planes. Hoek and Bray (1974) provided tables for estimating the values of A and B, and also gave more complex equations for estimating the effects of water pressure and cohesion on the joints. The usefulness of these simple charts is illustrated by a stability analysis of a wedge of sandstone perched high up on the cliffs of the Castle, a spectacular mesa south of Sydney (Fig. 3.10). Notwithstanding the steep dips of the bounding joints, and the seemingly precarious position of the wedge, substitution of values from Hoek and Bray's charts indicates that the wedge is quite stable, an estimate which is supported by the lack of any freshly broken surfaces on this section of the cliff.

Rotation and gliding of blocks

So far, we have considered mainly failures entirely within sandstones, but many failures which incorporate sandstones are seated in underlying shales or clays. Superb examples can be seen in the numerous fossil landslides along the Msak Mallat and Hamadat Manghini Escarpment of central Libya (Grunert and Busche, 1980). The northern part of the escarpment, which is cut mainly in Nubian Sandstone, rises steeply to a height of about 300 m above the adjacent foreland. It is generally capped by vertical cliffs; its lower slopes are intensely dissected by streams; and there is little evidence of major landslides. In the southern part, the escarpment morphology changes, and the steep slopes below the northern cliffs are replaced by an uninterrupted chain of landslide deposits. These deposits, which form a belt up to 3 km wide, consist of individual rotational failures more than 1 km long, over 100 m high, and about 200 m wide at the base. The failures have occurred in the thick red clays of the Tilemsin Formation; but each of them has incorporated a section of the overlying Nubian Sandstone which now lies as a broken, inclined capping on the slumped mass. The style of failure changes where most of the escarpment is cut in clays. Instead of being incorporated into

Fig. 3.10 The 150-m cliff on the western wall of The Castle, Clyde Valley, southern Sydney Basin. Note the intersecting joints on the wedge of sandstone (A); despite the precarious appearance of this wedge, estimates of geometric and frictional parameters indicate that it is stable. Note also the vertical alignment of the caverns, indicating the prominent role of water seeping through the rock mass in cave formation

rotational failures, blocks of the now thin Nubian Sandstone have tilted outwards as they have been carried downslope on mudflows. The blocks moved by these flows have disintegrated on the slopes.

Grunert and Busche emphasized that the failures along this escarpment did not occur on thin, highly plastic layers at the top of impermeable clays, but that thick masses of clay had behaved almost as a viscous fluid flowing downslope and across the adjacent foreland for several hundred metres. The failures are no longer active under the prevailing aridity, even during intense storms, and an annual rainfall at least a magnitude greater than the present 20 mm would be needed to saturate the clays (Grunert and Busche, 1980).

Margielewski and Urban (2003) describe a distinctive type of *sackung* or sagging failure of sandstone overlying shale in the Carparthian Mountains of southern Poland. Shearing stress resulting from repeated disturbance of slope equilibrium is unloaded along pre-existing joints or along newly developed fractures in the sandstone. As energy is thus absorbed within the slope, the initial movement involving the widening of joints is prolonged. Numerous crevice-type caves develop as a result of the slow movement of the rock, and the lack of slickensides indicates that crevice widening precedes the vertical or pivotal shift of blocks along the crevices. The consequent spreading of the sandstone mass increases pressure on the underlying shale, causing it to sag or creep. The sandstone blocks then begin to move vertically in response to disturbance of the shale. Once the shear stress limits are exceeded, catastrophic collapse occurs. Radiocarbon dating has shown that massive landslides were more frequent during humid periods, but the landslides were the final stage in a slow sequence of movement initiated by widening of cracks in the sandstone.

The last mode of failure to be considered here is block gliding, whereby blocks move outwards from clifflines while remaining intact and almost upright. Probably the first examples in sandstones to be described in detail are the 'Toreva blocks' of the Hopi Indian Reservation in Arizona, documented by Reiche (1937). These large blocks moved away from the main sandstone cliffs across failure planes in, or at the contact with, the underlying shales. Many of these blocks have tilted backwards as they moved downslope. More impressive examples occur on the eastern escarpment of the Chuska Mountains of New Mexico (Watson and Wright, 1963), where enormous blocks of the Chuska Sandstone are to be found up to 12 km from the present mountain front and several hundred metres below it. The contact between the sandstone and the underlying shale seems the likely plane on which horizontal movement occurred. But at least part of the horizontal movement and certainly most of the vertical displacement can apparently be attributed to the internal deformation of the unconsolidated and rounded sands beneath the indurated caprock, especially during periods when water tables were high (Watson and Wright, 1963).

Failure on underlying shales was also the cause of gliding of blocks in the Millstone Grit at Alport Castles in the Peak district of the English Pennines (Johnson and Vaughan, 1983). The morphology of the failure indicates that there were two phases of movement, in which an initial rock creep on the lower slopes induced high stressing and subsequent rotational gliding of the main blocks from the cliffline. Age determinations on a lens of peat incorporated in the toe of the failure indicate that the main movements occurred some time after 8300 BP.

Cambering (valley bulging) involving the downslope movement of limestone and weakly cemented sandstone is well documented in the Northampton Ironstone Field of northern England. The dip of the caprock into the adjacent valleys was first attributed to folding, but detailed borehole data showed that it was the result of slope displacement. When erosion penetrated these relatively resistant beds, the far less competent clay beneath them was squeezed out into the valleys by the pressure of the overburden. Downslope movement in this case seems also to have been enhanced by periglacial action.

Perhaps the most extensive area of large-scale block gliding in sandstones extends for about 25 km along the eastern side of the Cataract Canyon of the Colorado River. The displacement of blocks has produced a series of spectacular, roughly symmetrical, graben-like depressions in the sandstones of the Cutler, Rico and Hermosa Formations. Graben widths are generally from 150–200 m, and average depths are from 25–75 m. These rocks dip at about 4° towards a section of the canyon where the Colorado has cut down into ductile evaporites beneath the sandstones. The removal of confining stresses by the incision of the canyon has allowed the evaporites to plastically deform down-dip, causing fracturing and collapse in the overlying brittle rocks. The graben faults are apparently initiated close to the contact of the brittle and ductile rocks, and propagated both upwards and downwards. In response to local stress fields, they have developed a curved planimetric geometry, concave towards the canyon (McGill and Stromquist, 1979; Laity, 1987).

Another type of block gliding, in which very large blocks move almost horizontally outwards from a cliff at inclinations far less than the appropriate angles of friction, is less easily explained. Excellent examples of this sort are to be seen in the Nowra Sandstone south of Sydney, especially near the town of Nowra (Young, 1983b; Young and White, 1994; Young et al., 1995). That most of these blocks have moved outwards from adjacent cliffs, and have not been isolated by erosion, is beyond doubt – there are no streams feeding into the crevasses which now lie behind them; projecting plates on one wall of a crevasse fit neatly into recesses on the other wall; and in places, a slab of sandstone still bridges the crevasse (Fig. 3.11). At some sites, blocks 20–40 m high now lie between 100 m and 400 m out from the main cliffline. While some have fallen, most have moved down planes inclined at about 2–4°, and in a few instances, the lateral movement has been horizontal. The strong lateral stresses measured in the region south of Sydney readily account for the opening of joints, but they do not explain how the blocks could move so far. The angle of friction on highly plastic clay bands interbedded with sandstones can fall as low as 4°, and even to 0°, while the cohesion on their surface can range from 0–0.6 MPa (Jaggar, 1978b; Richards et al., 1981). But no such highly plastic beds lie beneath the sandstones incorporated in gliding

Fig. 3.11 An early stage of block gliding near Nowra, southern Sydney Basin. The caprock is still in place above the outer block, which has moved away and tilted back slightly. The central crevasse extends for another 40 m along intersecting joints

failures south of Sydney. On the contrary, the angle of friction at the contact between the sandstones and the underlying sandy siltstones and silty sandstones can hardly be less than about 30° (Jaggar, 1978b). Neither can the gliding be attributed to high porewater pressure, for the sites are well drained and lie within a few hundred metres of gently sloping interfluves.

The answer seems to lie in rheological deformation (Carey, 1953) by very slow creep of the underlying siltstone. This conclusion is supported by the difference in failure mechanism between coastal escarpment on the down-dip side of the plateau and the gorges further inland (Young and White, 1994). Where the escarpment cliffs face in the down-dip direction, block gliding is common, and the blocks have moved out across gently sloping benches cut in the underlying siltstones. On cliffs lining the gorges across the plateau, the dip of the strata is into the valley side. Here, the blocks topple when undercut, and the slopes on siltstones cut across the dip of the strata are 25–30°. Strain rates estimated for the siltstones vary with the thickness of deforming strata, and range from 270 m/Ma to only about 11 m/Ma (Young *et al.*, 1995). Any topographic expression of such slow rates of deformation would, under most circumstances, be obliterated by more rapid processes of erosion. Yet these estimates seem reasonable given the slow rates of long-term denudation determined from the excellent K–Ar chronology for the widespread basalts in this region (Wellman and McDougall, 1974; Young and McDougall, 1985).

Complex failures

Although the potential movement of blocks has been considered here as a series of distinctive modes of failure, it is obvious that an individual collapse may involve more than one type of failure. Thus, depending on the slope, the angle of friction and the width-to-height ratio, a block may fail by simultaneously toppling and sliding. The sliding may occur at the base, or it may take place along bedding or cross-bedding planes within the block. The very detailed account given by Schumm and Chorley (1964) of the fall of Threatening Rock in Chaco Canyon shows clearly how the collapse of cliffs can involve several processes of failure. Huge blocks of Cliff House Sandstone had become detached from the walls of the canyon by gliding outwards over the shales and thin inter-bedded sandstones and coal of the Menefee Formation. Dating of logs wedged beneath Threatening Rock to stop it collapsing onto an adjacent pueblo indicate that it was already threatening to fall at about 1000 CE, and back-calculation from measurements of the accelerating movement suggest that the rock began moving around about 550 BCE. The gliding was slow, for Threatening Rock did not fall until 1941, probably some 2500 years after it started moving. Photos and measurements taken prior to the fall showed that about 30 m of sandstone rested on about 5 m of shale exposed at the base, and that sections of the front of the detached block had been undercut almost to the axis of gravity, both by basal weathering of the Cliff House Sandstone and by erosion of the shale. Whereas movement prior to the fall had been dominantly a sliding away from the cliffline rather than a tilting, eye witness accounts state that the final movement was one of sliding and toppling.

Another example of compound failure can be seen south of Sydney (Young *et al.*, 1995), for example, at the Cathedral Cave on the northern bank of the Shoalhaven River at Nowra (Young, 1983b). At that site, a 20-m high block of Nowra Sandstone has moved downwards by about 7 m and outwards at its base by about 15 m. The upper part of the block now leans against the cliff wall, forming a triangular cave – 30 m long and 8 m high – above a rubble fill. The motion of the block was rotational in the horizontal, as well as the vertical plane, for the cave narrows from a mouth 6 m wide to end in a rock-choked cleft. As the block moved outwards, bedding planes on the block were tilted backwards 40° from the

Fig. 3.12 Chimneystack Rock at Yalwal, southern Sydney Basin. This towering block of Nowra Sandstone has moved outwards, gliding over underlying silt-stone for about 100 m from the nearby cliff

horizontal. Sliding then occurred along the bedding planes within the block as the inclination exceeded the angle of friction on these planes. Complete disintegration of the block by sliding and by secondary reverse toppling was only prevented by the closeness of the main cliff face. The sliding sections came to rest against the cliff before they slid off into the cavity behind the displaced block.

While complex failures may be common, the significance of thresholds between different types of failures cannot be ignored. This is made clear in the case of Chimneystack Rock, another very large detached block of the Nowra Sandstone. The Chimneystack (Fig. 3.12), which is 9 m thick at the base and 2 m thick at the crest, rises some 30 m from the surrounding sandstone rubble. Despite its slender and seemingly precarious form, field evidence leaves little doubt that it has moved outwards about 100 m from the adjacent cliffs (Young, 1983b). Yet toppling analysis shows that even such a slender pinnacle (with a width/height ratio of 1:12) would only begin to fall when tilted to about 10°; and at present, it is inclined at only about 4°. This inclination is too gentle to overcome the roughness between the individual beds within the block and allow sliding to occur. Movement is therefore limited to the base of the pinnacle, presumably at the sandstone–siltstone contact, and, as suggested previously, it is probably occurring as an extremely slow creep.

Whether the movement of an individual block triggers a more extensive collapse will depend on the function of that block in the total array of blocks. In some instances, the blocks will not be able to move because of their shape and orientation, or because they are held in place by neighbouring blocks. In other instances, once one block moves, neighbouring blocks which were previously restrained can move too.

Rock mass strength

The concepts already discussed are brought together in schemes to estimate the strength of rock masses. One of the most widely used is that of Hoek and Brown (1980), developed originally in relation to underground excavations, and revised to take account of wider applications (Hoek and Brown, 1997). The scheme brings together:

- the maximum and minimum effective stresses at failure;
- the uniaxial compressive strength of intact rock;
- a constant, m_i, derived from the relationship between effective stresses and the uniaxial strength of the intact rock, modified according to a Geological Strength Index (GSI) which varies with both structure and surface quality;
- two constants, s and a, related to the GSI.

The scheme provides field estimates of parameters other than the effective stresses. A scale from 0–6 places sandstones generally at grade 4 – strong rocks with a uniaxial compressive strength of 50–100 MPa, indicated by needing more than one blow of a geological hammer to fracture the rock. Estimates of the constant m_i are given for different rock types with sandstones and greywackes having similar values to dolerite and andesite, but lower than quartzite and considerably lower than granite or gneiss. The GSI is estimated from a chart where structure varies from interlocked blocky undisturbed rock to heavily broken rock, and surface conditions from fresh and unweathered to slickensided and clayey.

In New Zealand greywackes, Read *et al.* (2000) found that the Hoek–Brown criteria needed to be modified. The m_i value determined by laboratory testing was

Fig. 3.13 Sheer cliffs rise above sloping cut-rock slopes in sandstone at Keep River National Park, Western Australia. The joint orientation shapes the cliffs; and the faces indurated by iron oxide contrast with the faces on exposed softer pale-coloured rock etched by pits and runnels

50% lower than that estimated using the generalized scheme, and the accuracy of predictions was inconsistent, because the GSI did not adequately take account of defect spacings in the greywackes. The response however was to refine the parameters, not to reject the scheme. While the Hoek–Brown scheme is designed for, and used in, large-scale excavations such as dam sites or major tunnels, the principles are obviously applicable to erosional 'excavations' like cliffs and rock hillslopes.

Selby (1980, 1982), in his analysis of hillslopes formed on hard rock, looked first at slopes which fail along the outward-dipping joints (the term 'joint' including here joints, bedding planes, foliation or other discontinuities) (Fig. 3.13). As discussed earlier in this chapter, the failures are determined by the shearing stresses and friction along the joints. Then he reviewed slopes which are in approximate equilibrium between the resistance of the rock mass and the inclination of the hillslope, which he called strength equilibrium slopes. These exclude slopes which are:

• recently formed and thus without open joints, such as walls of fresh lava;
• slopes over-steepened by active undercutting;

- dominated by solutional erosion;
- relict, still maintaining a form developed under past conditions;
- entirely mantled by regolith;
- stabilized by buttressing at the base, although these may come into equilibrium as cross-joints open.

For strength equilibrium slopes, the important parameters are intact rock strength (measured by Schmidt hammer), weathering, spacing of joints, joint orientations, width of joints, continuity of joints and groundwater outflow. Bare rock slopes were found to fall within a range of angles for differing rock mass strength ratings. Very strong quartzites were in equilibrium at more than 65°, whereas strong sandstones stood at 35–70°.

Evolution of cliffs

Although these techniques for predicting the general strength or failure criteria for clifflines are obviously very useful, they need to be considered with reference to activities on the footslope. As Selby (1982) pointed out, buttressing can support a steeper-than-equilibrium slope. The analysis of slopes developed on sandstone in Antarctica carried out by Augustinus and Selby (1990) showed that, in positions high on a slope, sandstones form steep faces which are essentially controlled by the mass strength of the outcrop. Low on the slopes, however, the same rock lies at much gentler angles, similar to the angle of repose of their thin talus mantles. These lower slopes are Richter slopes, where the angle of the underlying bedrock adjusts to the angle of the overlying regolith mantle.

Similar phenomena have been described from the very different environment of the Orange Free State in South Africa (Munro-Perry, 1990). Where the Clarens Sandstone acts as a caprock, slopes develop by parallel retreat, with the sandstone forming vertical cliffs above talus-covered slopes of the Richter type inclined at 30–35°. However, where the entire section is cut in the sandstone, a second phase of slope replacement occurs in which the Richter slopes are progressively replaced by low-angled surfaces inclined at only about 14°.

The preservation of steep cliffs in many instances depends on basal erosion of weaker strata beneath a sandstone. The variation in morphology along the Arnhemland Escarpment of northern Australia is a case in point. Galloway (1976) has shown that this escarpment has prominent vertical cliffs, cut in the highly resistant Kombolgie Sandstone, only where substantial thicknesses of weathered schists or granite crop out on the lower slopes. Where the weaker rocks no longer crop out and the Kombolgie Sandstone extends down to the fringing pediments, the escarpment is much less sharply defined and has rounded rather than vertical faces. Prolonged stabilization of the erosional edge of the Kombolgie Sandstone results in a very weakly defined and dissected escarpment above pediments that extend onto the sandstone (Galloway, 1976). This clear-cut link between the preservation of vertical faces and basal sapping on the footslopes is not because the sandstone capping the escarpment is weak. On the contrary, the Kombolgie Sandstone is highly quartzose and strongly cemented.

Variations in the form of escarpments may be controlled by major bedding planes within the sandstone caprock, rather than by changes in the exposure of weaker rocks on footslopes (Oberlander, 1977). Oberlander pointed out that changes in slope form in massive sandstones are usually associated with partings in otherwise uniform rock. The rock below the parting – for example, a major bedding plane – acts as a surrogate for weaker substrates like those described by Galloway for the Arnhemland Escarpment and by Koons (1955) for sections of the Colorado Plateau. Lithology and dip angles are also important. As Gregory and Moore (1931) pointed out, particular ridge-forming sandstones on the Colorado Plateau tend to display an individual set of minor topographic characteristics. Cretaceous sandstones commonly make sharp-angled hogbacks, frequently with a narrow, serrated comb along the crest; the massive Navajo Sandstone forms a lofty ridge with rounded summit outlines, and back slopes deeply grooved by lateral canyons; the Wingate Sandstone forms compact, sharp-angled ridges with largely undissected crests and dip slopes.

Much emphasis has been given in the geomorphological literature to the accumulation of talus at the foot of cliffs. A popular view may be that cliffs rise above slopes covered by the blocks that have fallen from them, but in fact, deep talus accumulations below cliffs are probably the exception rather than the norm. Often it is argued that, if talus is not removed, the lower section of the cliff will become progressively buried by it, resulting in a progressive decline of bedrock slopes to the angle of repose of the debris mantle (Young, 1972). There is good evidence to suggest that armouring of the footslope may result in a substantial decline of the slope of the cliffed segment or to a slowing of the retreat of that segment. Koons (1955) proposed a stop-and-go sequence of retreat for cliffs of the arid southwest USA, with active retreat during phases of basal erosion separated by phases of stability which resulted from the armouring of footslopes by talus. Furthermore, the general preservation of vertical faces on the cliffs of that region, even to a final stage of isolated pinnacles such as those of Monument Valley, has been attributed to the absence of widespread, thick talus accumulations (Gregory, 1917, 1938; Schumm and Chorley, 1966). The lack of talus may be due locally to slow rates of retreat of cliff faces, but on a regional basis seems to be much more the result of a rapid disintegration of talus (Schumm and Chorley, 1966). Gregory (1917, p. 130) gives a memorable summation of the importance of the paucity of talus mantles:

Much of the material supplied to the base of the cliff is fine, the product of disintegration of poorly compacted sandstone. In the areas of the Wingate, Navajo, and de Chelly sandstones large blocks pried off from the tops of cliffs are so lacking in firmness that they crumble to sand on striking the lower slopes; the conditions for the formation of talus are absent. Cliffs in this region are steep not because the rocks are hard but because they are friable. . . As in general talus is the regulator of cliff profiles, the action of erosive forces which prevent its persistent accumulation is in large part the controlling cause of the mural aspect of the scenery.

Similar conclusions have been drawn from the arid sandstone terrain of southern Jordan (Goudie *et al.*, 2002). Although there are many scars of rock falls in this landscape, the scarcity of talus is attributed to the mechanical weakness of these sandstones, which disintegrate when they hit the slopes beneath.

In the southwest USA, it seems to be the presence or absence of thick talus which really determines whether slopes can be considered as being in equilibrium with the rock mass strength or whether they are controlled by footslope processes. This point can be illustrated by reference to two sandstone towers (see Gregory, 1917, Plate VI) – Mitten Butte, in Monument Valley, is cut from the massive and strongly jointed De Chelly Sandstone; whereas Organ Rock, in Moonlight Valley, Utah, is cut from jointed, though very closely bedded shales and sandstones of the Moenkopi Formation. Notwithstanding the great contrast in lithology and structure, both of these penultimate erosional remnants rise vertically from footslopes that have only thin and discontinuous veneers of talus, and appear to be controlled largely by footslope processes.

Mills (1981) described the variable effects of different types of sandstone debris on hillsides in the Appalachians. He observed that whereas thinly bedded sand-stones break down in small fragments, thickly bedded sandstones – such as the Tuscarora Sandstone – break into large boulders. These boulders move down-slope and concentrate in depressions on the underlying slopes cut in shales. The armouring provided by the boulders diverts erosion to adjacent outcrops of shale which are denuded and leave the armoured section as higher ground. This sequence of events is similar to that described for the incision of pediments mantled with coarse debris in the semi-arid Flinders Ranges of South Australia (Twidale, 1967).

The importance of the relationship between talus and cliff development can be illustrated further by turning again to the examples of the sandstone landscapes of the Eastern Kimberleys. As described previously, the Glass Hill Sandstone of the Bungle Bungle Range undergoes granular disintegration, but has a compressive strength capable of maintaining steep slopes. Moreover, the granular debris is readily swept away across pediments at the base of outcrops, leaving largely bare rock footslopes from which towers and knife-edged ridges rise abruptly (see Figs 1.4 and 1.5). The relationships are displayed even more clearly on the slopes

Fig. 3.14 In the Cockburn Range, Western Australia, cliffs stand at the heads of streams cutting back into the escarpment. Where talus mantles the ridges between, the cliffs are degraded

of the Cockburn Range. Where the sandstone talus mantle has been stripped from the underlying shales by streams draining down the sides of the mesas, the cliffs are high and nearly vertical; but where the talus has accumulated on the ridges between the streams, the cliff slopes have generally degraded into a series of small steps and risers. From a distance, these degraded slopes can be mistaken for thick talus accumulations reaching almost to the cliff tops, but they are cut-rock features mantled by talus (Fig. 3.14). A mass strength rating (see Selby, 1982) of the more closely bedded and jointed sections of the Cockburn Sandstone yields a predicted equilibrium slope of about 40°, which matches closely the declivities of these sections. The changes from bold clifflines to degraded slopes in these sandstones in places occur over a distance of less than 100 m across the face of the mesas. Cliffs are limited to those outcrops in which fractures are widely spaced, and, especially, to outcrops below which the talus mantle has been stripped, allowing basal sapping in the underlying shales.

Amphitheatres

The evolution of cliffs needs to be considered in the horizontal plane, as well as the vertical. Embayments that extend horizontally in cliff walls are a striking feature of most sandstone country. Some of these embayments are remarkably angular and have been termed *rincons*, 'the Spanish word for the inside corner of a house' (Gregory, 1917, p. 132). Joint control seems the obvious cause of their angularity, but Gregory believed that they are primarily the product of the disintegration of the rock by weathering and by thin sheets of water flowing down the face of the cliff, aided by groundwater seepage. He emphasized that they should be distinguished from semi-circular recesses and from box canyons by their lack of well-defined drainage lines.

Not all embayments formed primarily by the disintegration of rock faces are angular. In a very detailed study of landforms in the Organ Rock Formation of the eastern Colorado Plateau, Nicholas and Dixon (1986) have attributed small arcuate embayments to the preferential breakdown of rock faces in zones of high frequency of joints. Even in the very strongly jointed sandstones of the southern Sydney Basin, where cliff retreat is primarily a process of block detachment, most embayments take the form of beautifully arcuate amphitheatres. This is also the most common form of embayments cut in massive sandstones as a result of large-scale basal sapping at points of concentrated seepage of groundwater (Howard and Kochel, 1988; Laity, 1988). The erosional processes by which embayments are initiated and enlarged are considered more fully in later chapters; our concern here is with the arcuate plan form.

An amphitheatre is roughly analogous to an arch lying on its side, and so any lateral stresses will tend to hold individual blocks on the curved face in place. This form is more stable than a straight wall. It is also more stable than a narrow, angular slot, in which high concentrations of stress may promote accelerated failure (Philbrick, 1970). The analogy between amphitheatres and arches in the horizontal plane is particularly apt for areas like the southern Sydney Basin, where the major and intermediate principal stresses are essentially horizontal. A selection of the planimetric forms that are typical of amphitheatres in this region is shown in Fig. 3.15. Most of them are elliptical, and we have found only one that is essentially semi-circular. Because of the dominantly elliptical shape their orientation, in relation to the principal regional stresses, is an important determinant of the patterns of boundary stress around them. Figure 3.15, which matches those elliptical forms to the diagrams presented by Hoek and Brown, gives a very general indication of those stress patterns. We draw attention particularly to the great contrasts between the stresses around those with similar forms, but with markedly different orientation in relation to regional stresses. We emphasize that the relationships shown are no more than general guides, for the actual patterns may, in some instances, be affected considerably by stress shadows created by adjacent canyons. It should also be noted here that the arcuate forms of amphitheatres can be

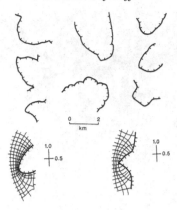

Fig. 3.15 Sketches of the planimetry of eight amphitheatres cut in sandstone cliffs in the Shoalhaven valley in the Sydney Basin. The lower two diagrams show the generalized distribution of stresses around different types of amphitheatres, showing the relative magnitude of horizontal stresses typically found in this region

distorted considerably by stream erosion along major joint sets intersecting the face.

The variations in stress on a rock face must also be considered in three dimensions. The greater stability of planimetrically curved faces has been demonstrated clearly in field and experimental studies of open-cut mines. Early work by Pitteau and Jennings (1970) quantified the relationship between planimetric curvature and slope stability in the deep open-cuts of South African mines, and found that the distance of slope failure, or breakback, from the edge of originally vertical shafts varied with the radius of curvature at the foot of the failure. Hoek and Bray (1974) likened the effect of planimetric curvature to very flat wedge failures. As the angle of the wedge decreases, the horizontal component becomes increasingly important in supporting the wedge, and gives rise to 'arching' with a significant increase in stability. These authors suggested that where the radius of curvature of a concave slope is less than the height of the slope, the angle of the slope may be 10° steeper than that calculated by two-dimensional analysis, but when the curvature is convex, the slope may be 10° flatter than calculated by two-dimensional analysis.

The relationship between height, planimetric curvature, and slope inclination and stability was demonstrated experimentally by Stacey (1974). The slopes of his models of jointed, inclined blocks were most stable when the radius of curvature was equal to the height, and became increasingly less stable as the planimetric curvature was flattened. His models also showed why there was a marked decrease in stability from a tightly curved face to a straight wall. The movement of blocks in a curved face was clearly inhibited by their neighbours.

Fig. 3.16 A waterfall drops more than 200 m over a broad amphitheatre in the Danxia landscape at Dabaiyan, Chishui, southern China. (Photo: Mr Hong Kaidi)

This effect could be seen not only in the horizontal plane, but also in the vertical. On the tightly curved face, blocks were displaced individually. As the curve flattened increasingly, deep crevasses formed when whole towers of blocks failed – first by sliding and then by toppling – until finally a stage of catastrophic failure across the entire face occurred when the radius was about four times greater than the height. Nonetheless, Stacey estimated that the stabilizing constraints of curvature continue to be felt until the radius is about 20 times greater than the height.

The implications for sandstone geomorphology of these studies of open-cut mines are obvious, but direct comparisons are made difficult by the uncertainties of field measurements. The main difficulty that we have encountered is that of accurately measuring the radius at the base, for unlike the quite youthful mine pits, natural amphitheatres have generally undergone considerable erosional modification (Fig. 3.16). Average slopes are also often difficult to estimate because of abrupt changes at major lithological boundaries. In searching for a natural analogue to the relationships demonstrated in mines, we have compared the radius of curvature along the crest of sandstone cliffs to the local relief of the amphitheatre. Pilot studies of amphitheatres near the southern end of the Sydney Basin have yielded generally consistent results. The frequency distribution of different sizes of amphitheatres ($N = 85$) shows that 72% have a radius ranging from 200 m to 500 m, and that 90% have a radius of less than 700 m. The relationship between radius of curvature and height of the amphitheatre over a considerable range of sizes was consistent. A set of 60 amphitheatres in the sandstone lands of the Shoalhaven River gave a correlation coefficient of $+0.82$; and for some sub-catchments, the coefficient rose to $+0.92$. What is more, 75% of the total sample of 85 amphitheatres had a radius-to-height ratio of less than 4, and 90% had a ratio of less than 5. These clear trends in the field measurements indicate that the dimensions of the amphitheatres are far from random, and suggest that there is a three-dimensional equilibrium form approximated by many amphitheatres.

The engineering studies of open-cut pits just summarized attempt to predict the magnitude of slope failure, given initial morphologies defined in terms of radius of the curved footslope and its ratio to the height of the wall. We have considered morphologies developed at some time subsequent to the initial state, and contend that the general relationships between form and stress distribution, which constrained development from the initial state, still apply. Given that the curvature from the centre of the amphitheatre to the top of the cliffs is equivalent to the initial floor radius plus the breakback distance, as defined by Pitteau and Jennings, the consistency of our results surely points to the operation of the same general principles. And it is important to note that all of the radius-to-height ratios listed by us fall well within the critical limit of 20:1 cited by Stacey as the approximate limit of the effect of planimetric curvature. Indeed, almost all of our examples fall below a ratio of 5:1, with approximately 20% of them below 2:1. Bearing in mind that many of the cases measured by us may have been deepened by erosion as well as expanded by the retreat of their walls, the tendency to maintain an approximately equilibrium form which is linked to the minimization of three-dimensional stress concentration seems to have constrained the morphological development of the amphitheatres.

In this chapter we have considered cliffs in sandstones in terms of brittle materials responding to the forces acting upon them. Yet brittle fracture and block collapse or gliding, even with the added effects of underlying weaker rocks and of talus mantles, are not the only influences on cliff morphology. Nor are rectilinear slopes the only major type of morphology developed on sandstones. Bare rock can have high curved faces, or even quite gentle slopes, and it is to these forms which we now turn.

4

Curved slopes

Though the Navajo stands first among cliff makers in the plateau province, it does not form platforms or mesa tops ... Its composition, texture, and structure combine to produce smooth or ribbed mounds on which streamways are poorly defined ... (there is) a maze of domes and saucer like depressions.

... the massive, strongly cross-bedded Navajo sandstone cliffs are straight or undercut at the base and rounded at the top, and surfaces between canyons weather into domes, haystacks and tepees.

(Gregory, 1938, p. 10)

Explaining curvilinear forms such as these, which are by no means limited to the Navajo Sandstone, remains a major task. This task is made difficult by the interaction of three broad sets of controls, one set being structural, the second being lithological, and the third being erosional (Fig. 4.1). The difficulty is increased by the variation of the curved forms; these range from domes (also known as pinnacles, turrets, beehives or pagodas) and smooth 'slickrock' slopes, through alcoves, to arches and natural bridges. Variations in scale must also be considered, as small curved features, especially polygonal tessellation and 'elephant skin', are frequently superimposed on larger curved surfaces.

Cross-bedding

For Gregory (1917, 1938), the cause of the curved forms cut in the sandstones of the Colorado Plateau – especially, though not exclusively, the Navajo Sandstone – was self-evident:

To the tangential cross-bedding are due the exceptional erosional features of the Navajo sandstone – the innumerable pockets, recesses, and alcoves bounded by curved planes which characterize this formation. Overhanging cliffs are common, and the beautiful arc of the Rainbow Bridge is only an unusually perfect example of the control exerted by curved laminae.

(Gregory, 1917, p. 58)

Fig. 4.1 Domes above cliffs cut a strongly bedded sequence in the Entrada Sandstone, Colorado Plateau

This theme is emphasized again and again in Gregory's work, and at first sight it seems reasonable enough. The cross-bedding in the Navajo and in many of the other sandstones of that region truly is outstanding, and the control by the cross-bedding of small-scale relief is clearly evident on outcrop after outcrop. But whether this control applied equally to large-scale features was never really demonstrated by Gregory, and has since been disputed.

Bradley (1963) considered cross-bedding to be very much subordinate to large-scale jointing in constraining the curved outcrop of sandstones on the Colorado Plateau. Howard and Kochel (1988, pp. 8, 9) drew conclusions similar to Bradley's, noting that 'despite these microscale lithologic controls, the slickrock slopes generally show only minor form control by poorly jointed sandstone such as the Navajo Sandstone'. They supported this assertion with excellent photographic evidence of smooth slopes grading across abrupt changes in the dip of very prominent cross-bedding in the Navajo Sandstone in Zion National Park (Howard and Kochel, 1988, Figs 4, 7 and 8). We too have seen many instances of smooth surfaces transecting prominent cross-bedding in that region, especially in the valley of the Escalante, and not a single unequivocal example of cross-bedding control of the scale proposed by Gregory. Neither have we seen any convincing example of slope control at this scale in dominantly cross-bedded sandstones in Australia. On the contrary, probably the most striking examples of rounded sandstone terrain in Australia are in rocks dominated by horizontal bedding. The towers cut in the Glass Hill Sandstone of the Bungle Bungle Range (see Fig. 1.4) and those cut in the Sydney Basin sandstones (see Figs 4.5, 4.6) are cases in point.

Jointing

Stress-released fracturing of sandstone, that produced large curved sheeting roughly parallel to the topography, was seen by Bradley (1963) as the dominant control of the curved morphology of extensive sections of the Colorado Plateau. At many sites in that region, there is no doubt that sheeting (exfoliation) is the dominant control. Some of the most striking instances are in Zion Canyon, where curved fractures cutting across the bedding form 'a crude, somewhat subdued replica of the surface form' (Bradley, 1963, p. 520). One of the best examples in the Zion National Park is in the Kolob section (Fig. 4.2). There, great curved slabs are peeling from a magnificently rounded dome that could well pass, at least at a distance, for a typically rounded granitic inselberg (Fig. 4.2). Another fine example in the Zion area is the curved rock slab, looking like a flying buttress on a cathedral, that overlooks the Virgin Valley near Springdale (Gregory, 1950, Fig. 123). This slab of Navajo Sandstone is 47 m long, less than 2 m wide and

Fig. 4.2 Curved fractures control the form of a large dome in the Kolob section of Zion National Park, Colorado Plateau, USA

reaches a maximum height of about 16 m above the ground. Good examples can also be seen in Capitol Reef National Park, Utah, where curved sheeting – ranging in size from huge fractures in the Wingate Sandstone to slabs a few metres in length on domes in the Navajo Sandstone – cuts across the bedding. Indeed, curved fracturing is an outstanding feature of numerous sandstone outcrops over this region of southwestern USA. Nonetheless, whether such fracturing can be invoked to explain most curved topographic surfaces is questionable. Indeed, Howard and Kochel (1988) disagreed with Bradley's interpretation and argued that exfoliation or unloading is the exception rather than the rule on the smooth, often gently curved surfaces of slickrock slopes of the Colorado Plateau.

Large-scale sheeting is also prominent on the flanks of Uluru (Ayers Rock), the great monolith cut from arkosic sandstones in central Australia. The rock is extremely massive, and dips at about 80°, but has no prominent jointing other than the curved sheeting. Some sections of the flanks have a double curvature where the steeper upper convex slope gives way to a more gently curved footslope. Ollier and Tuddenham (1961) suggested that the general convexity, including the areas of double curvature, results from the generation of slab failure by unloading. They envisaged that a gentle footslope develops as the main wall retreats; and that when the footslope is large enough, it too begins to generate unloading shells which are more gently inclined than those on the face above. An alternate explanation of the double curvature is that the lower, flared slopes are exhumed features developed in a zone of intense weathering at the former junction of the inselberg and the surrounding lowland (Twidale, 1978). Ollier and Tuddenham certainly recognized weathering, specifically hydration, as the cause of accelerated small-scale spalling of the arkose along watercourses on the flanks of the inselberg. The extensive thin spalling on the ridges between the watercourses was less readily explained. While recognizing the contrary experimental evidence, Ollier and Tuddenham still invoked thermal expansion and contraction resulting from diurnal insolation. Nevertheless, it is unloading to which they attributed the dominant sheeting on the flanks of the Uluru inselberg.

Although large shells develop beneath many rock faces, whether they are due to the elastic response of rock to the erosional reduction of compressive stresses is still very much a point of contention. The first detailed proposal for the residual tectonic stresses being released by erosion was probably Gilbert's account of sheeting on granites in the Sierra Nevada (Gilbert, 1904). This concept has become a tenet of modern geomorphology, and is repeated in virtually every textbook on the subject. Thus fracturing of the rock triggered by the reduction of compressive stress has seemed the most ready explanation for the development of curved joints in sandstone. Twidale (1973), however, argued that many so-called

Fig. 4.3 Sheeting on the slopes of Uluru (Ayers Rock), central Australia. (Photo: C. R. Twidale)

unloading joints are actually joints formed by compressive stresses deep in the crust and that they were subsequently opened by stress adjustment to erosion (Fig. 4.3). Indeed, he pointed out that the occurrence of A-tent phenomena – in which shells of rock are prised upwards from the surface of some outcrops – is proof that many such curved fractures occur in rock masses which are still in compression. In cases such as these, he argued, the obvious conclusion is that the joints controlled the curved topography, not vice versa. This line of argument has been restated by Vidal Romani and Twidale (1999).

Yatsu (1988) disagreed with Twidale, arguing that curved joints are indeed a response to topography, but he nonetheless rejected the standard appeals to unloading. Such fractures, he maintained, are not formed by the release of residual compressive stress, but are the response to tensile stress set up during the excavation of valleys. In arguing that sheeting is a response to contemporary stresses on valley sides, Yatsu appealed to patterns of fractures mapped at a dam site near Hiroshima and to the finite element modelling by Kawamoto and Fujita (1968). He gave particular emphasis to Kawamoto and Fujita's prediction of extensive tensile zones on steep slopes (see also Stacey, 1970) as the most probable cause of sheeting. In arguing this case, he also drew attention to the predicted alignment of the minimal (tensile) principal stress roughly normal to the slope, implying that tensional sheeting will likewise tend to parallel the contemporary surface. As the modelling by Stacey (1970) predicted, and as field observation by Carlsson and Olsson (1982) demonstrated, the presence of substantial lateral tectonic stresses is the major determinant of the size and intensity of tensile zones on valley walls. Carlsson and Olsson emphasized that the effect of lateral stresses which substantially exceed local vertical stresses is to create an essentially uniaxial stress field in the horizontal plane. Moreover, as excavations in sandstone near Sydney have shown, horizontal stress fields close to the surface certainly can produce fractures through previously intact rock (Braybrooke, 1990).

An alternate approach to the problem of curved fractures is to estimate the direction in which they propagated by measuring the alignment of the patterns on their surfaces. The alignment of these curves indicates that many of these fractures expanded upwards, for they show vertical walls topped by single or multiple arcs. On closer inspection, it can be seen that many of these vertically curved fractures in sandstones originated above major discontinuities in bedding, and thus are very probably the product of tensile stresses developed on canyon walls. This seems true also of slabs that have failed laterally across overhanging faces.

Over and against these varied criticisms of the unloading theory is the clear evidence of the allied and well-known phenomenon of valley floor bulging. Unloading of residual compressive stress has seemed the only feasible way in which strata on valley floors, rather than valley walls, could be distorted. Such natural unloading is echoed by the valley bulging, cracking of bedrock floors and tilting of joint blocks caused by subsidence resulting from underground mining. Yet caution is required because, as is known from studies in open-cut mines, sidewall stresses in pits can produce buckling on the adjacent floor simply because of a concentration of shearing stresses adjacent to the footslope (Stacey, 1970).

Perhaps the debate about the origin of curved fractures may, in part, be due to the use of examples of landforms which have had diverse origins. Some fractures may have been formed at depth, and others near the surface. The debate may also have arisen from a conceptual difficulty, in that it cannot really be simplified into the dichotomy of 'push or pull', for a 'push' may well enhance a 'pull'. In short, while compressive stresses may be present at the regional scale, their main effect might be to create tensile fields on valley sides, and those tensile fields could result either in the opening of pre-existing joints or in the propagation of new fractures. It seems that there is no simple answer to the origin of curved fractures in sandstone that can be applied in all cases.

While the origin of such fractures is still debated, engineering studies have done much to explain the extent of rebound and fracture development (e.g. McQueen, 2000). The key factors determining it appear to be the magnitude of the initial stress field, the extent of subsequent excavation, and the relationship between the direction of maximum stress and long axis of excavation. Rock strength (as a proportion of stress magnitude) and rock mass structure (especially weakening elements such as shale bands), major bedding planes and joints are also significant. The removal of vertical and lateral constraint can release stored strain energy, resulting both in instantaneous and time-dependent rock movement. Moreover, as noted in the previous chapter, erosion can cause redistribution, as well as the release, of *in situ* stresses.

Critical partings

The walls of many valleys in sandstone change angle in the vertical plane, with gentle cut-rock footslopes giving way abruptly to steep cliffs. The abrupt change from one type to another and the shift in the boundary between them, in the Entrada Sandstone of The Arches National Park, Utah, have been explained in detail by Oberlander (1977). Oberlander attributed some of these changes, especially in sandstones of low permeability, to the wetting and consequent reduction in strength of rock above prominent partings near the crest of cliffs. However, he attributes the development of slickrock slopes from steep cliffs primarily to slab failure above 'effective' partings in the bedding, that is, above partings that are the sites of undercutting. These partings consist either of closely spaced bedding with highly fractured sandstone, or of fissile shales. Below the partings are slickrock slopes which extend at the foot of the retreating cliffs. Oberlander presented convincing evidence of the control of changes in the position of the break in slope by variations in the dip of the effective partings or by their pinching out (Fig. 4.4). Similar examples of the development of slickrock slopes below critical partings under cliffs occur in the Cowhole Sandstone of eastern Arizona, especially at the type locality of that sandstone (Harshbarger *et al.*, 1957, Fig. 33).

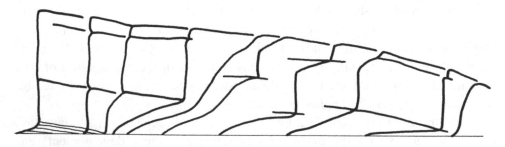

Fig. 4.4 The effects of critical partings in sandstone on the sequential development of cliff and slickrock slope segments (after Oberlander, 1977)

To return to a point made previously in Chapter 2, tensile zones on valley sides seem to be virtually eliminated as slope angles decrease below about 30°, especially where lateral tectonic stresses are low (Stacey, 1970). Consequently, as slickrock slopes develop below the undercuts along effective partings, the stress distribution over the entire canyon wall changes. As slickrock slopes become more prominent, there is less potential for tensile zones on the canyon wall. This means that the slickrock slopes buttress the canyon wall. And because tensile zones on them are minimal, the slickrock slopes themselves do not form sheet fractures and are stable. The stage in Oberlander's model where effective partings pinch out and the near-vertical slab wall stagnates has the most stable configuration with minimum concentration of stress.

Oberlander's model was designed to explain slickrock slopes in generally massive sandstones with only a few prominent horizontal bedding partings, but gently inclined and curved slopes can also develop in strongly bedded sandstones. In the latter case, however, the slopes are not smooth, but are broken by numerous small, convex risers between gently sloping benches. Striking examples of this type of terrain occur in Morton National Park in the southern Sydney Basin, where a multitude of bedding partings in the sandstones of the Berry Formation has been eroded into gently sloping contoured outcrops which appear as concentric rings on aerial views. Pickard and Jacobs (1983) argued that in this part of the Sydney Basin, such partings are accentuated by thin beds of softer sandstone. Our field evidence lends little support to this claim, and we believe that the breaks in slope are due mainly to variations in permeability and consequently of seepage through the sandstone itself. Where the partings are very closely spaced, the breaks in slope become muted, and the surface takes on a generally convex form (Fig. 4.5). These conclusions from the Sydney Basin are strongly supported by evidence from the Bungle Bungle Range, where sandstone in the recessed partings is generally indistinguishable from that on the adjacent slopes, except for apparent changes in permeability.

Weathering and erosion

Although structural constraints can apparently account for many curved surfaces, they by no means account for all of them, and perhaps not even for the majority. Even at localities where there is strong planar vertical fracturing, and the fracture planes clearly once extended beyond the modern confines of the outcrop, summits are generally rounded. Such a combination of prominent vertical jointing and rounded summits is well displayed in many parts of the Colorado Plateau, especially in Canyonlands National Park and in Glen Canyon near the mouth of the Dirty Devil River.

Fig. 4.5 Erosion and cavernous weathering of a joint-bounded 'pagoda' near Mudgee, western Sydney Basin, have produced multiple small convex slopes between bedding planes

Schumm and Chorley (1966) suggested that the main cause of rounding of highly permeable sandstones on the Colorado Plateau is weathering. Their experimental results showed that water enters most of these sandstones very readily, and they concluded that the rounded cores in the experimental blocks indicate how weathering eliminates sharp edges on natural outcrops. In the case of sandstones with high porosity but low permeability, like the massive Entrada Sandstone, thorough wetting of cliffs is likely to be limited to their upper part where water can seep down joints and concentrate near major bedding planes. Schumm and Chorley's experimental work is supported both by their own field evidence of weathering of these sandstones and by Gregory's emphasis, after decades of field work in that region, on the highly weathered nature of many of the sandstone surfaces (Gregory, 1917, 1938).

Fig. 4.6 Curved faces on cliffs in Monolith Valley, southern Sydney Basin, Australia, are cut by myriad near-parallel runnels. Note also the tubes and small caverns along several of the bedding planes

Howard and Kochel (1988) did not support structural origins for slickrock slopes and concluded that most slickrock slopes are essentially the result of grain-by-grain weathering and erosion, and of the peeling of thin (< 1 cm) weathered rinds. They pointed out that many slickrock surfaces on the Colorado Plateau exhibit prominent solution features such as weathering pits and solutionally deepened runnels. This is true also of similar sandstone surfaces in the Sydney Basin (Fig. 4.6). Moreover, Howard and Kochel's emphasis on the importance in the formation of smooth or gently curved slopes of the grain-by-grain erosion of highly weathered sandstone, and the detachment of very thin sheets of sand-stones, is very strongly supported by field evidence from the Bungle Bungle Range of Western Australia. Skins of weakly cemented grains can easily be detached from weathered faces; grains can easily be brushed by hand from freshly exposed surfaces; and most material transported down the flanks of the towers moves as individual grains.

The importance of clast-by-clast breakdown, albeit with some mass collapse, can also be demonstrated in terrain cut from conglomerate. The great domes of Kata Tjuta (the Olgas), in central Australia, are cut from extremely coarse conglomerate in which the diameter of individual clasts ranges from 25–40 cm (Fig. 4.7). Thick debris mantles of boulders and pebbles on lower slopes attest to

Fig. 4.7 Smoothly curved cliffs cut in conglomerate on Kata Tjuta (the Olgas).
Thick debris mantles lie at the foot. (Photo: G. Nanson)

a clast-by-clast breakdown, although, as Ollier and Tuddenham (1961) noted,
sheeting parallel to the curved slopes occurs on many outcrops. The rounded
spires of the Ragged Range in northwestern Australia are no less impressive than
those of Kata Tjuta. These too are cut from an extremely coarse conglomerate,
and the thick debris mantles, which in places are burying the outcrop, again attest
to a clast-by-clast breakdown. There is no sign of sheeting structures in these
conglomerates. The rounded towers cut from conglomerate at Montserrat in
Spain are another example. Where cementing between clasts is strong, slopes fail
in slabs and the imprint of jointing constraints is more pronounced. For example,
the massive Kialas conglomerate of southern Tibet has in places been shaped into
superbly rounded towers and domed summits which, on many slopes, parallel
very large curved sheeting structures (Gansser, 1983, Plate 31). Similarly, at
Meteora in Greece, the influence of jointing is obvious.

    Although smoothly graded surfaces on friable material are generally explained
in terms of the critical slope for transportation of debris, Howard and Kochel
(1988) consider slickrock slopes on the Colorado Plateau to be 'weathering-
limited', that is to say, the potential rate of transport is much higher than the rate
at which debris is produced. Slopes on the towers of the Bungle Bungle Range
also seem to be controlled largely by the resistance to detachment of grains. The
angles of the mid-slopes of the towers are remarkably consistent around a mean

of 64°, an inclination that is consistent with the high internal angle of friction of these closely interlocking grains (R. W. Young, 1987).

The complex origins of curved forms

In some cases, such as the famous central Australian inselberg, Uluru (Ayers Rock), curved slopes seem to be the result of a combination of the factors considered above. Bremer (1965) interpreted its beautifully curved form as the product of weathering under a humid tropical climate that prevailed during the Tertiary. She envisaged climatic conditions, and a general regime of deep and intensive weathering, akin to those which shaped the great sugarloaf forms of coastal Brazil. Whether or not the processes that shaped those crystalline outcrops necessarily were the same as those that shaped the arkosic sandstone of Uluru remains conjectural. Be that as it may, in support of her assertion, Bremer argued that the numerous grooved channels on the flanks of Uluru were formed by solution, not by abrasive gullying; and also that the lateritized erosion surface of the surrounding lowland was evidence of a tropical weathering regime. Bremer interpreted the large tafoni (cavernous weathering) on the flanks as probably the result of weathering under warm semi-humid conditions, and suggested that the basal 10–15 m of the slopes are multi-period features resulting from weathering during humid times and stripping during drier ones. She certainly recognized the great sheeting structures as 'azonal' or structural features, but considered them as of secondary importance in producing the curvature of the slopes. Pollen recovered from sediments in fossil drainage lines on the adjacent lowland (Twidale and Harris, 1977) and elsewhere in central Australia certainly leaves no doubt that the area did have a humid climate during much of Tertiary times. Indeed between 90 million and 15 million years ago, much of inland Australia was forested under warm humid climates; and 50 million years ago, extensive freshwater lakes in the Lake Eyre basin supported fish, crocodiles, freshwater dolphins and turtles (Johnson, 2004). Uluru has been shaped slowly during this long span of time extending back to the Early Tertiary (Twidale, 1978; Twidale and Campbell, 2005). Yet the evidence for a climatically induced rounding is essentially circumstantial, and must be weighed against the observations by Ollier and Tuddenham (1961) and Twidale (1978) which point to the dominance of structural and lithological control of slope forms.

Twidale (1978) considered that, as well as unloading joints, basal weathering and the growth of caverns on the footslopes have caused the convexity of the slopes. Since rock ribs and channels extend over the full height of Uluru, and numerous solution holes and pools occur from the midslopes to the crest, the surface is being actively changed (Fig. 4.8). The inselberg is certainly not simply

Fig. 4.8 Cavernous weathering cuts into the base of slopes on Uluru, central Australia. Large runnels, weathering pits and pools extend from the midslopes to the crest, indicating active erosion of the inselberg. (Photo: C. R. Twidale)

a relict feature. Major sheet structures produced by compressional unloading of the rock mass maintain the general curvature of the slopes. One factor that has received surprisingly little attention in the explanation of the curved form is that this Cambrian sandstone was tightly folded during the Devonian, and has a near-vertical dip. Consequently, the sandstone did not form a caprock, and weathering and erosion extended down the bedding planes over the full height of the inselberg. Hence the arkose sandstone is weathered and susceptible to granular disintegration. Also the weathering has penetrated down near-vertical joints that very roughly parallel the unloading joints. The significance of the dip is made clearer by a comparison with the typical mesa morphology of Mt Connor, 50 km to the east, where the same sandstone is flat-lying and does act as a caprock. Clearly both Mt Connor and Uluru have developed under the same long-term climatic and geological forces, so the difference in form is due to the difference in the ways those forces have operated, as dictated by the dip of the beds.

Apses, arches and natural bridges

The dominance of curvature in many sandstone landscapes is not limited to domes or slickrock slopes, but is also obvious in the form of recesses developed by mass failure or erosion into the faces of cliffs. Gregory (1917) has described the arch as the dominating architectural form for the Colorado Plateau. There are, however, important variations from this dominating architectural form. Except for some sections of the Colorado Plateau, true arches and natural bridges, with broad openings extending right through a bedrock wall, are relatively rare. Apses, alcoves or blind windows, in which there is a bedrock backwall to the curved structure, are much more common. One could hardly improve on the beautiful description of variations in these forms written by Gregory (1938, p. 105):

Some of the arches appear as if drawn on the smooth rock face; others form the borders of panels and blind windows recessed in the cliff faces; still others serve as the roofs of niches and rock shelters. As erosion continues, most of the arched structures are detached from the walls, and lacking adequate support, collapse in response to gravity, exposing new curves and arches behind them.

We follow Gregory's lead in considering the development of these varied arcuate forms, turning first to the origins of the gaps in the rock, and then to their morphological stability.

Apses (alcoves)

The simplest to explain are curved recesses created by the collapse of curved joints or by the upward propagation of slab failures of elliptical form. A mass of rock falls from the cliff wall, leaving the roof of the recess bounded by the arcuate form of the fracture or joint. For example, about half of the opening under the Skyline Arch, in The Arches National Park, Utah, was created by a single rock fall in 1940 (Schumm and Chorley, 1966). Excellent examples of various stages in the development of curved forms created essentially by mass collapse can be seen in the Wingate Sandstone in Capitol Reef National Park and in Glen Canyon, Utah (Fig. 4.9).

Weathering of cliff faces – especially weathering caused by seepage of groundwater from the cliff – also plays a major role in initiating the openings around which arched forms develop. This certainly was Gregory's opinion, for he argued that 'in early life they express the work of groundwater' (Gregory, 1938, p. 105). Furthermore, he suggested that the role of seepage often increases in importance as the recess deepens and more surface water makes its way down through fractures in the arched roof (Gregory and Moore, 1931). Gregory's emphasis on groundwater has received much support in recent years, especially from studies of the Navajo Sandstone (Laity and Malin, 1985; Howard and Kochel, 1988; Laity, 1988); these studies are considered in detail in Chapter 6.

Fig. 4.9 In Capitol Reef National Park, Utah, curved fractures cut across the bedding in Wingate Sandstone, and slabs break away from the rounded face

Arches and natural bridges

Arches and natural bridges occur in many sandstone areas (Fig. 4.10). Worldwide, there are about 400 longer than 50 m, with Landscape Arch in The Arches National Park being the longest at 88.3 m. The highest, at about 365 m, is Tushuk Tash (Shipton's Arch) in western China. While arches occur in other lithologies, these impressive titleholders are in sandstone. Certainly, of the 10 longest natural arches in the world, all are in sandstones. Seven are in Utah–Arizona and only one is found outside the USA (the Aloba Arch in the Ennedi Range in Chad) (The Natural Arch and Bridge Society, 2008).

Possibly the greatest concentration of natural sandstone arches is in and around the Tassili National Park in Algeria, where more than 490 arches of varying sizes have been recorded (Debossens, 2007). Blair (1984) noted that there are more than 300 natural arches in southeastern Utah, and attributed this high concentration to the properties of the Entrada and Cedar Mesa Sandstones in which most of these arches are found. In The Arches National Park, the Entrada Sandstone is cut by long, closely spaced joints that have been widened by erosion, leaving numerous thin rock fins. Arches form when the fins are breached. He suggested that the breaching may be facilitated by the high content of readily leached carbonate and clay cement of the sandstones.

Fig. 4.10 Owachomo Bridge in Natural Bridges National Monument, Utah

Natural bridges present an interesting variation on the theme of seepage, for here it is not regional movement of groundwater, but seepage of water from stream channels through the narrow walls of incised meanders that creates the opening. Again we are indebted to Gregory's pioneer research, especially on the bridges of White Canyon, in Natural Bridges National Monument, Utah. He suggested that the growth of meanders incised in the Cedar Mesa Sandstone, in which White Canyon is cut, eventually created bedrock walls sufficiently narrow for water to seep across the necks of meanders (Gregory, 1938). He cited the contact of massive sandstones and shaley interbeds as the preferential pathway for seepage. The seepage created 'underground runways' that were progressively enlarged into a split through which the White River was diverted, creating an ever-enlarging gap under the sandstone girder of the bridge, and leaving the apex of the meander abandoned. The field evidence at the bridges in White Canyon, and at the renowned Rainbow Bridge, in a small tributary canyon on the southern side of the Colorado, leaves no doubt of the soundness of Gregory's interpretation. The largest sandstone arch in Europe, the 16-m high and 27-m long Pravcicka Brana arch in the Czech Republic, has also developed by lateral erosion through a narrow rock rib (Varilova, 2002).

How then do the arches and bridges opened in this way subsequently develop into apparently very stable forms? Useful analogies can be drawn from studies of the stability of cavities in underground mines. The experimental studies by Stephansson (1971) are of particular interest, because they demonstrate that the pattern of fractures in the roof of cavities depends largely on the ratio between the thickness of major beds and the height of the cavity. When this ratio was low – with the thickness of the layers less than half the height of the roof of the cavity – applied loads produced vertical to subvertical fissures in areas of high bending moments, that is, in the centre of the roof and above the abutments. When the thickness of the individual layers exceeded half the height of the roof, a very different type of failure took place. Failure in these thick layers was initiated by a vertical crack developing in the centre of the roof and extending upwards to about the zone of zero normal stress along the neutral axis between the compressed and tensile parts of the layer. Simultaneously, or with only a very small increase in load, two additional cracks propagated towards the centre of the neutral axis from the edge of the abutments. Collapse of material from these two fractures produced an arched roof. The arching has been explained as a failure due to higher shear stress in the centre of the beam in relation to the normal stress at the bottom edge, but Stephansson argued that it is dependent on fractures propagating towards the centre of the neutral axis. Close agreement between the behaviour of the models developed by Stephansson and the observed failure and configuration of the roof of the Swedish Kautsky mine in horizontally bedded sandstones and conglomerates increases confidence in the application of these experimental results to natural arches.

The first approximations of tangential stress in estimating the potential for stress-induced failure in rock surrounding tunnels can also be applied to the development of natural arches. Tangential stress is given by the equation

$$\sigma_T = 6\sigma_H - \sigma_V \qquad (4.1)$$

where σ_T is induced tangential stress, σ_H is horizontal stress and σ_V is vertical stress (McQueen, 2000). The extent to which tangential stress approaches UCS of the sandstone indicates the likelihood of failure. Failure induced by tangential compressive stress is indicated by V-shaped shear fracturing in the roof. This mode of failure is particularly significant in areas such as the Sydney Basin, where there are substantial tectonically induced horizontal stresses.

It should be remembered that the roof of a natural cavity on a cliff may undergo slow spalling long before it collapses. In 1991, a piece of rock 22 m long fell off Landscape Arch, and another 14 m long dropped in 1995 (The Natural Arch and Bridge Society, 2008). Robinson (1970) argued, using beam analysis, that stress patterns in a flat roof are such that there is a tendency for minor failures or spalling to result in concave forms, because the underside of the bed forming the roof is in tension, and because the stress trajectories through that bed are concave upwards. Robinson also suggested that stress patterns around a more fully developed arch are more accurately approximated by the stress patterns around a circular opening than by stresses in a beam. Again, however, the rock above the centre of the arch is in tension, and again the stress trajectories are arcuate. Hoek and Brown (1980) provided stress trajectory diagrams for a considerable range of shapes of cavities and for different ratios of principal applied stresses. They drew a comparison between stress trajectories and streamline patterns that is useful for the intuitive interpretation of the diagrams; they liken zones of tension in an elastic material to streamline separation, and zones of compression to streamline crowding.

Hoek and Brown also pointed out that the ratio between the applied tensile and compressive stresses is a major determinant of the pattern of the stress field around a cavity. This ratio is expressed as the coefficient of applied stresses (K):

$$K = \frac{\text{Average horizontal stress}}{\text{Average vertical stress}} \tag{4.2}$$

Hoek and Brown recognized a series of critical thresholds with reference to the value of K. When $K = 1$, the optimum shape of the cavity with lowest boundary stresses is a circle; for other values of K, the optimum shape is an ovaloid opening in which the axis ratios match the stress ratios. For circular cavities, they also noted that when $K = 0$, boundary stresses in the roof and floor are tensile; when $K = 0.33$, stresses in the floor and roof are zero; and when $K > 0.33$, all boundary stresses are compressive. The effects of variations among these factors can readily be seen in the diagrams of stress trajectories and ratios between major principal stresses given by Hoek and Brown.

Hoek and Brown (1980) used the analogy between stress trajectories and streamlines around the piers of a bridge to describe stresses around multiple cavities. If the piers are aligned across the direction of flow, the streamlines are compressed around the piers with zones of undisturbed flow between the piers. The analogous zones of distortion of stress around a circular cavity extend a distance of about three radii from the centre of the cavity. If the piers are parallel to the flow, slack water zones develop between them. The analogous situation is a zone of reduced stress, or stress shadow, between the cavities. Clearly, these concepts are important in considering series of neighbouring arches or recesses in a cliff face. They also need to be kept in mind when considering the effects of valleys or sequences of mesas in regions where horizontal stresses are dominant.

Although the general distribution of stress around arches and caves provides a guide to their development, the specifically structural properties of these features must also be considered. The roofs of many natural sandstone cavities resemble a flat lintel, as one or more flat-lying blocks bridge the gap beneath. The weight of the blocks acts as a vertical load, bending the blocks and inducing horizontal stresses that are compressional on the upper side and tensional underneath. The blocks may fracture, but if the broken pieces are not dislodged, they can remain in place, continuing to transmit the vertical load into a horizontal thrust onto the abutments.

The truly curved arch is generally more efficient because the vertical load is transmitted as if through a curved line of blocks. The load pushes the blocks against one another. This keeps the blocks in compression and transmits the thrust to the abutments. If we extend the argument on the stability of blocks outlined in the previous chapter, it can be seen that, provided the thrust line of the transmitted load stays within the curvature of the arch, and especially within the central third of the arch, the blocks will not move, and the structure is likely to remain stable (Heyman, 1982). There are multiple positions that curved thrust lines can occupy within a given arch (Heyman, 1982). If the arch becomes progressively thinner, the number of possible positions decreases, until only a single stable curve remains. Yet, as long as a single thrust line can be found for the complete arch which is in equilibrium with the external loading, and which lies entirely within the masonry of the arch ring, then the arch is safe. Further thinning creates instability, with the thrust line extending outside the arch, and this leads to the hinging of blocks along the arch. Yet, even if the arch should fracture, it will still not collapse unless at least four hinge points develop between the blocks, allowing the structure to fold up on itself. Indeed, if an adjacent pair of the four hinges opens in the same direction, the arch may still remain stable, as the pair will act as if it were a single split hinge (Heyman, 1982).

As horizontal stress is transmitted to the abutments, they too must remain stable if the arch is to stand. Additional vertical loading of the abutments by rock masses rising above the footing of the arch will reduce the eccentricity of the thrust line, much like the effect of statues or pinnacles atop the abutment of the flying buttresses of cathedrals. The shape of the arch also affects the stability of the buttresses – the flatter the arch, the greater are the horizontal stresses that it generates.

The natural arch differs in one essential way from its engineered counterpart. Unlike the radially aligned, wedge-shaped blocks of an engineered arch, the blocks of a natural arch will generally be aligned in one direction which transects the arched face for at least part of its length. In other words, the faces between blocks will not all be aligned perpendicular to the thrust, so that a consequent generation of shearing stresses can tend to displace the blocks. Nonetheless, natural arches are remarkably stable, especially when cut from massive sandstones that have significant tensile strength. Yet fail they do, as the remnants of collapsed arches along White Canyon, described by Gregory (1938), prove.

The principles outlined above can readily be applied to the morphology of natural arches and bridges. The Rainbow Bridge in Utah is carved from the

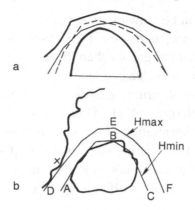

Fig. 4.11 Thrust lines of sandstone arches. (a) Rainbow Bridge, Utah (95 m high) showing the geometric centre line (dashed) and the corresponding thrust line (funicular polygon). (b) An arch in the Colo Valley near Sydney. This arch will remain stable while its thrust line stays within the range defined by polygons for maximum (H_{max}) and minimum (H_{min}) horizontal stress configurations

Navajo Sandstone. Its beautiful parabolic curvature, the thickening of the bridge from 13 m on the crest to more than 25 m approximately midway up the supporting spans, and its termination in massive abutments, make this an extremely stable natural arch. Its stability can be illustrated by comparing its form as estimated from photographs with the approximate thrust line calculated by a funicular diagram (Heyman, 1982). The thrust line shown here closely follows the centre line of the arch (Fig. 4.11). As the deviations between the two lines are small, the thrust line could be contained within an arch of the same shape that was only about half as thick as Rainbow Bridge, provided the thinning of the left span was from the top rather than from below. Even if erosion from below placed this section of the thrust line outside the arch, only a single hinge-point would develop and the structure would still be stable. In fact, as the arch is cut from massive sandstone, the tensile strength of the rock would probably retard the opening of even this hinge.

Natural arches and bridges do have quite diverse geometries. Owachomo Bridge, in Natural Bridges National Park, has a long, flat central girder which spans 55 m, but is only about 3 m thick (Gregory, 1938) (see Fig. 4.10). The load is thus transmitted horizontally across the girder to the abutments, rather than obliquely down an arch. As Gregory has argued from the amount of incision of the channel since it was diverted to form the bridge, Owachomo probably represents a late stage in the evolution of natural bridges, in which substantial collapses have reduced the thickness of the sandstone girder. Yet flat arches can be quite stable, provided the abutments can carry the horizontal thrust and provided the dimensions and jointing of the girder offer little scope for hinging.

Similar conclusions can be drawn for the long, flat, thin girder of the natural bridge near Doyle's Creek, northwest of Sydney (see Fig. 4.11). The gap spanned is more than 10 times greater than the average thickness of the girder. The stability of this natural bridge can be estimated by applying standard formulae for the maximum shearing stresses and tensile stresses generated in supported cantilevers (Herget, 1988). Although the rectangular shape of the opening ensures high concentration of stress at the junction of the girder and abutments, estimates of the shearing stresses are well within the range of the shearing strengths of these sandstones. However, the tensile stress generated in the centre of the girder is about 4 MPa, a figure close to the tensile limits of these sandstones.

Polygonal cracking

Curved slopes on sandstones may be decorated by polygonal cracking, commonly called 'elephant skin weathering'. This form of tessellation should not be confused with patterns found on benches or platforms due to weathering and widening of intersecting joints. Many of the best examples of this latter type of patterning occur on the wave-swept, flat surfaces of shore platforms. Such patterns of intersecting fractures on shore platforms may be accentuated by case hardening caused by precipitation of thin vertical bands of silica and iron oxide along the faces of the joints (Fig. 4.12). Indurated patterns of tessellation may also be caused by Liesegang diffusion through intact sandstone between joints. Nonetheless, the more rapid seepage of water along joints often constrains the configuration of Liesegang shells, so that they conform to the joints. Very regular patterns of polygonal jointing in sandstone sometimes occur adjacent to igneous dykes. These instances seem to be the result of shattering or thermal metamorphism during intrusion of the igneous rock, a notable example being near Bondi Beach in Sydney (Branagan, 2000).

However, most forms of tessellation are not related to jointing, being surface phenomena that die out within a few centimetres depth, and follow the curvature of outcrops, even appearing on overhanging faces (Fig. 4.13). Usually, the pattern consists of polygonal plates that are essentially convex upwards. In places, the convexity has been so accentuated – presumably by weathering and incision along the edges of the plates – that the surface resembles a tightly spaced array of small domes (sometimes called 'tray of scones weathering'). Branagan (1983) noted that the surfaces on which many plates develop – as distinct from the individual plates themselves – may be flat or concave upwards, but the plates are by no means limited to gently sloping surfaces. He also noted that the variable size of the plates seems to depend on the dominant size of grains in the sandstone; and their geometry is in some places controlled by master fractures extending along the edges of a whole series of plates.

Fig. 4.12 Induration of closely spaced intersecting joints by iron oxides forms a complex boxwork on a rock platform cut in muddy sandstone, southern New South Wales

Fig. 4.13 Tessellation plates cover the curved surface of a Hawkesbury Sandstone outcrop. Note the finer tessellation on the steeper faces, suggesting some stretching under the influence of gravity

The best-known examples are from the vicinity of the Elephant Rock at Fontainebleau, France (Termier and Termier, 1963; Doignon, 1973; Robinson and Williams, 1989). At Fontainebleau, the long axis of the plates is drawn out to form roughly parallel bands (Williams and Robinson, 1989). The average size of the polygons is 100–200 mm in diameter; some are as small as 10 mm; and the plates are generally pentagonal (Williams and Robinson, 1989). Good examples have also been reported from near Boulder, Colorado (Netoff, 1971) and Zion National Park (Howard and Kochel, 1988). The list of examples from Australia given by Branagan (1983) includes several parts of the Sydney Basin, Keniff Cave in central Queensland and sites near Hobart, Tasmania. Other excellent Australian examples are in the Grampian Ranges of Victoria, the Sydney Basin and the Bungle Bungle and Kununurra areas of the northwestern part of the continent. Although very well-developed examples such as those listed here are infrequent, the phenomenon in general seems to be widespread, occurring on many massive, medium-grained and fine-grained sandstones.

The possible effects of climate on the development of tessellation are by no means clear. At first sight, the morphological similarities to polygonal structures of periglacial origin in poorly consolidated material raises the possibility of the polygons on sandstones also being some sort of freeze–thaw phenomenon. But this hypothesis offers no explanation of cracking on vertical faces, and is very much at odds with the apparently widespread occurrence of 'elephant skin' on sandstones in temperate areas. Williams and Robinson (1989) noted that it is rarely found in the humid tropics or the high arctic, and suggested that it is most abundant where annual rainfall is low or where there is at least a well-marked dry season. However, the examples in the Sydney Basin occur where rainfall occurs in all seasons, and on average exceeds 1000 mm per year. Branagan (2000) has suggested that most tessellation in this same area is probably the result of weathering during cold periods in the Pleistocene, but evidence from upland swamps on the sandstones of this area indicates that no major climatic change occurred during the Late Pleistocene or Early Holocene (A. R. M. Young, 1986).

There has been considerable debate as to the origins of this type of cracking. Termier and Termier (1963) thought that it might be a diagenetic feature, but this idea can be discounted because of the way in which the pattern follows the local variations of outcrop, and because it clearly dies out just under the surface of the blocks. In the Grampian Ranges of Victoria, for example, collapse of joint-bounded blocks has revealed that polygonal fracture patterns on even the most extremely tessellated faces die out within about a metre of the surface and the depth is generally much less. As the sandstone on which Netoff (1971) observed polygonal cracking is clay-rich, he proposed that the cracking might well be a desiccation phenomena related to variable clay content and composition. This mechanism could not, however, explain the excellent examples at Fontainebleau and near Sydney which are on sandstones with a low clay content (Branagan, 1983). Chan *et al.* (2007) proposed that the development of polygonal cracking in the Navajo Sandstone is a weathering feature controlled predominantly by tensile stresses determined by properties such as changes from massive to laminated beds. They suggested also that different forms and sizes of cracking may be controlled by age or stage of development.

A striking feature of the Fontainebleau and the Sydney Basin examples is a surface crust or rind formed by the concentration of secondary silica as a cement and as coatings on quartz grains (Branagan, 1983; Robinson and Williams, 1989; Williams and Robinson, 1989). Branagan reported that Netoff's sites in Colorado also display such a rind or skin, and the sites that we have seen in northern Australia have skins several millimetres thick composed of clay and secondary silica. Robinson and Williams (1989) suggested that the polygonal cracking thus seems to be the result of shrinkage of silica gel due to changes in temperature or

Fig. 4.14 Roughly pentagonal cracking separates dimpled raised tessellation plates on a platform in Hawkesbury Sandstone, Royal National Park, Sydney Basin

moisture during or after the formation of the crust. A detailed study by Robinson and Williams (1992) in arid periglacial conditions of the High Atlas in Morocco again emphasized the importance of surface crusting in the development of tessellation. They argued that it develops as the crust of silica, iron and manganese on the surface of the sandstone becomes denser and more rigid, and cannot adjust to changing stress by small intergranular movements.

Branagan (1983) suggested that the cracking of such skins is due to fatigue and changing surface stresses. His argument was based on a comparison between the silica-rich skins of the polygonal plates on sandstones and the glaze, or fused layer of glass on ceramic surfaces. Cracking or crazing occurs in glaze because of surface strain and because the glaze contracts more than the material below it. When first applied, the glaze has sufficient strength to resist surface strain; but as it ages, and stress conditions change, fatigue cracks are initiated and subsequently propagate to form an irregular network. Branagan suggested that a similar chain of events occurs in the silica-rich skins, although he noted that the often-regular patterns of cracks on sandstone more closely resemble the cracks in photographic emulsion on old glass plates, than they do the irregular crazing on pottery. Adjustment to surface strain, a time-dependent reduction of strength and the consequent development of fatigue cracks, seem to offer the most plausible explanation for the way in which the cracking closely follows the surface of many outcrops (Fig. 4.14). This sequence also explains why tessellation is not found on all case-hardened surfaces. Once the cracking has started, water concentrated in the slight indentations can accelerate weathering there. Another unexplained aspect of tessellation is the development of pockmarks in the centre of plates at some sites. Again perhaps, it may be due to stress distribution – this time within the individual plates – creating a point of weakness which is then exploited by weathering processes.

5

Chemical weathering

The physical properties and processes considered so far exert on, and in turn are influenced by, chemical and biological weathering. For example, stress relief promotes weathering because it opens up a rock mass to the ingress of water. This effect is most obvious as joints open, but can be just as important at a very much smaller scale if it results in extensive micro-fracturing. Such fracturing increases the surface area available for weathering. For example, a 1-m^3 block fragmented into 10-mm cubes increases the surface area by 10^6. Dry clays previously locked within the rock can draw in atmospheric moisture and aerosols. This in turn can trigger further breakdown, by slaking, owing to the pressure exerted by capillary suction. Indeed, where expansive clays are present, the pore spaces may explode as a result of suction pressure that may be of the order of 100 MPa (McNally, 1993). Even where there is little expansive clay, initially slightly weathered rock can become loose and flaky within a few weeks of exposure. Moreover, micro-fracturing resulting from stress relief can rapidly increase the susceptibility of sandstone to other processes of weathering. McNally (1993) reported that in a 60-m deep road excavation in the Terrigal Formation sandstone, north of Sydney, Australia, fresh strong sandstone suffered a 30% loss in the sodium sulphate soundness test which measures susceptibility to the effect of salt. Porosity and saturated water content increased, and strength and elastic modulus decreased. Although this quartz–lithic arenite in its unweathered state has compressive strengths of 60–100 MPa, when extremely weathered its strength is < 5 MPa. So great has been the effect of prolonged and intense weathering that fresh rock in the Terrigal Formation is now only encountered in tunnels or in excavations 50 m or more deep.

This example is by no means exceptional, and the once widespread belief that quartz-rich sandstone is virtually chemically inert (see, for example, Tricart and Cailleux, 1972) must be abandoned. Estimates of relative chemical erosion rates for major rock types (Meybeck, 1987) place sandstone with mica schist and granite (rates relative to granite 1–1.3) below volcanics and shale (1.5–2.5) and

113

most metamorphics (5), and well below carbonates (12) and evaporites (40–80). Yet, while erosion rates may be slow, long periods of weathering and/or intense weathering in hot and humid areas can produce fine examples of true karstic terrain (see Chapter 6).

However, it is not always easy to distinguish the effects of near-surface weathering from those of diagenesis that took place at depth. This is shown by research into the weathering of quartzite on the old Palaeic surface of the Varanger Peninsula in northern Norway (Fellanger and Nystuen, 2007). Weathering along zones of fracturing in these quartzites has significantly weakened their mechanical strength; so that clastic quartz grains as well as quartz overgrowths (which are covered by very thin clay coatings) are easily loosened, and are the source of sheets of friable sand along mountain sides. The presence of small amounts of kaolinite in the weathered zones was first interpreted as the result of intensive weathering during Cainozoic or Mesozoic times. But the subsequent identification of illite and pyrophyllite indicated a more complex mineralogical history for the quartzite. It now seems that kaolinite or some form of mixed-layer clay filled pore spaces and coated grains of clastic quartz during deposition in NeoProterozoic times. With increasing depth and temperature during the early burial of the sediments, much of the kaolinite was transformed into illite; and it was also at this stage that quartz dissolution and re-cementation occurred as overgrowths took place. Deeper burial, with temperatures reaching around 300 °C, resulted in the formation of pyrophyllite at the expense of kaolinite and quartz. The strong corrosion of quartz in contact with pyrophyllite certainly testifies to the disintegration of quartz at that stage. Weathering on the Palaeic surface involved mainly the penetration of groundwater down fractures and along grain contacts with clay coatings, resulting in the weakening of the rock. Chemical processes in this weathering event were limited largely to the mobilization of iron from the clays and its subsequent precipitation as Fe^{3+}, although loss of potassium from the illite may have led to the neoformation of small amounts of kaolinite (Fellanger and Nystuen, 2007).

The chemical breakdown of sandstone by near-surface weathering is to a considerable extent the result of the removal of the more soluble components of the intergranular cement and matrix. This is certainly the case where carbonate is a major component, and the role of organic material and dissolved CO_2 in the production of acidic water that dissolves the carbonate is well known. Examples of calcareous sandstones include the now classic study by Frye and Swineford (1947) of karst forms developed in the Cretaceous sandstones of central Kansas, and the observation by Howard and Kochel (1988) of the development of solution pits on the Colorado Plateau by the solution of calcite. Discharges of orange-coloured groundwater demonstrate the removal of iron

Fig. 5.1 Solutional weathering along joints and bedding planes, and within the sandstone blocks, has deeply and intensely etched this highly quartzose sandstone shore platform at Jervis Bay, New South Wales

from many sandstones. Usually, the iron has been mobilized as Fe^{2+} under anoxic conditions within the rock mass, and oxidized and re-precipitated as Fe^{3+} as groundwater emerges from an outcrop. However, it is the importance of the removal of silica, both from the cement and from the grains, that has been highlighted by the study of the weathering of sandstone in recent decades (Fig. 5.1).

Arenization and the formation of karst in quartzose sandstones

In carbonate rocks, 80% of rock bulk may be removed by dissolution, but the figure for quartzose landforms is much lower, perhaps 10–20% (Martini, 1979). However, Jennings (1983, p. 21) argued that solution must be 'critical (but not necessarily dominant) in the development of the landforms and drainage characteristic of karst'. Martini (1979) coined the term 'arenization' for the interaction of processes leading to karst development in quartzose rocks:

- Slow chemical dissolution of quartz occurs, especially along crystal boundaries.
- This frees individual grains, and the rock becomes less coherent and more susceptible to physical erosion.

- Sand grains are mechanically removed.
- A plentiful flow of water in the vadose zone leads to piping and removal of rock bulk.

Thus while processes other than true solution largely form the landscapes, 'solution plays an essential precursor or "trigger" role' (Ford, 1980, p. 345).

The thermodynamics and – it would seem – especially the complicated reaction kinetics are critical in the formation of silica karst. Martini (2000a) argued that the rate of reaction is just as important as the total solubility – the faster the rate, the shorter the distance that solutions can penetrate the rock before saturation. This results in arenization close to the surface and a general surface lowering, rather than a deep karstification. Slower rates allow joint widening without surface lowering; and slower still is crystal boundary solution with a deep solution of the rock. Voids along crystal boundaries are very thin, and thus water circulation is very sluggish, with saturation being reached after a very short distance unless the kinetics of reaction are very slow. Martini argued that if the rate of silica solution were slower, without changing the total solubility, karst on quartzite would be much more common.

Piccini (1995) also noted that, in the Roraima, arenized rock breaks down to sand and is removed by flowing water. There, arenization acts most effectively along the joints, where circulating water has a greater time of reaction between water and rock than water flowing over the surface. On the surface, runoff waters do not have enough time to dissolve the silica cement of the arenite; and the alternation of wet and dry conditions actually leads to the formation of hard crusts of silica cement and iron oxides which protect the rock surface. He found that the development of a karst-type landscape has been possible because the environmental conditions have limited the effects of mechanical weathering, allowing – over a very long time – the strong development of solution forms. Mechanical processes are active, but their effects are mostly concentrated along the streams, especially near the border of the plateau, and inside the active caves. Doerr (1999) has listed a combination of factors that have allowed the intense karstification in Venezuela:

- Whilst the unweathered rock may be of very low porosity, a dense network of joints and micro-fissures has allowed water to infiltrate and initiate corrosion.
- The arenite is very pure. In rocks of a more heterogeneous mineralogy, fissures enlarged by corrosion may become clogged by weathering residue.
- High precipitation counters the low total solubility of silica.
- There is little soil or other sediment which could block passages, preventing further karstification.
- No other geomorphic processes such as glaciation, periglaciation, submergence under the sea or frost weathering have interfered with karstification.

- Very long periods of subaerial exposure to weathering with stable geomorphic processes are required for karstic development, and corrosion for at least tens of millions of years.

.

Silica solubility and chemical kinetics

Much investigation into the solubility of silica in water has been based upon thermodynamic principles. It must, however, be noted that although 'thermodynamics is a powerful tool for elucidation (of) geological phenomena where equilibrium is normally attained ... many geological processes are controlled by reaction rates so that they can only be understood in terms of kinetics. Such a situation is illustrated by reactions at low temperatures in the silica–water system' (Rimstidt and Barnes, 1980, pp. 1683–1684).

Thus the kinetics of the reaction, or the rate of the solution process, may be just as important – or even more important – than the total solubility of the reacting species. Reaction rates between silicates and fluids near the Earth's surface are slow enough for fluids to be often out of equilibrium (Brady and Walther, 1990). Martini (1979) also emphasized that water may flow off a surface without having achieved saturation so the kinetics of dissolution, more than the solubility itself, may often be the most important consideration.

The solubility of silica under natural conditions is low compared with that of carbonates; the rate of chemical weathering of quartzose rocks is therefore much slower than that of many other rocks. Compounds of silica are present in all natural waters, either as suspended solids, as colloids, or in solution (Aston, 1983). Yet most silica is released into the environment by the weathering of silicate minerals such as feldspars, micas and clays, and very little comes by direct dissolution of quartz (Henderson, 1982).

Slow rates of chemical weathering on quartzose rocks can be expected because the equilibrium solubility of quartz is low, generally less than 30 mg/L (ppm) (0.5 mmol/L). Compare this with the solubility of limestone which may be several hundred milligrams per litre, depending upon temperature and pCO_2 (Jennings, 1985).

In its simplest form, the congruent dissolution of silica can be written (Henderson, 1982; Brady and Walther, 1990; Martini, 2000a) as:

$$2H_2O + SiO_{2(qtz)} \rightarrow Si(OH)_4^0$$
$$2\log K = -3.7(25\,^{\circ}C) \tag{5.1}$$
$$K \approx a_{H_4SiO_4(aq)} = 1.1 \times 10^{-4}$$

This $Si(OH)_4$ monomer is the main form of dissolved silica (Iller, 1979), and under near-neutral pH conditions generally exists as uncharged silicic acid,

H_4SiO_4. In natural aqueous solutions this silicic acid monomer tends to join or polymerize with other $Si(OH)_4$ units to form hydrophilic acids $(Si_nO_2)_{n-m}(OH)_{2m}$ or hydrophobic amorphous silica $SiO_2.xH_2O$ (Yariv and Cross, 1979).

If the silicic acid dehydrates partially, it forms polymers such as:

$$2H_4SiO_4(aq) \Leftrightarrow H_6Si_2O_7(aq) + H_2O \qquad (5.2)$$

Martini (2000a) notes, however, that the formation of this polymer is negligible in solutions undersaturated with respect to quartz. It only becomes dominant in strongly supersaturated quartz solutions. The monomer naturally dissociates ionically to:

$$H_4SiO_4(aq) \Leftrightarrow H_3SiO_4^- + H^+ \qquad (5.3)$$

However, the weathering of silicate minerals usually leads not to congruent, but to incongruent dissolution, i.e. it involves the formation/precipitation of new solid phases. The weathering of potassium feldspar, for example, results in the formation of silicic acid in solution and kaolinite clay (Aston, 1983):

$$4KAlSi_3O_8 + 22H_2O \rightarrow 4K^+ + 4OH^- + Al_4Si_4O_{10}(OH)_8 + 8H_4SiO_4(aq) \qquad (5.4)$$

The weathering of sodium feldspar may lead to precipitation of either aluminium hydroxide or kaolinite. In either case, silica is released in aqueous solution as uncharged monosilicic acid, H_4SiO_4. However it will not necessarily remain in solution, but may precipitate as amorphous silica or opal-A, or participate in neoformation of clays (Velbel, 1985). These processes clearly are reversible – the opal may re-dissolve or the clays may weather. Furthermore, over time, diagenesis can transform opal to more ordered forms of silica, to cristobalite–tridymite, thence chalcedony and finally quartz (see Wray, 1999). Thus, particularly in saprolite and soils, silica is present not only as quartz but also as amorphous and poorly crystallized forms. These last are far more soluble than quartz itself.

The forms and solubility of naturally occurring pure silica

Pure silica occurs naturally as eight distinct forms. Five of these show crystalline structure (quartz, tridymite, cristobalite, coesite and stishovite) and three are amorphous (amorphous silica, opal-A and lechatelierite) (Krauskopf, 1956; Jones and Segnit, 1971; Yariv and Cross, 1979). The abundance of the eight allotropes of silica in the geologic environment is not equal. Lechatelierite is a silica glass and very rare, as are coesite and stishovite which are found only in meteorite craters (Yariv and Cross, 1979). The other five silica polymorphs are quite common, with quartz probably being the most abundant. In the study of sedimentary rocks, the silica species of real interest are quartz, opal-A and amorphous silica (Siever, 1962).

Discussion of the solubility of silica must clearly distinguish between the different silica polymorphs, for they do not have the same solubility (Siever, 1962); and indeed silica is 'unique in the enormous difference in solubility among its polymorphs' (Krauskopf, 1956, p. 5). According to Krauskopf (1956), Siever (1962) and Yariv and Cross (1979), the solubility of amorphous silica ranges from 60–80 mg/L (ppm) (1–1.3 mmol/L) at 0 °C, to 100–140 mg/L (1.7–2.3 mmol/L) at 25 °C, and about 300–380 mg/L at 90 °C for most pH levels. Quartz is less soluble, with its equilibrium solubility at normal Earth surface temperatures being 6–14 mg/L (Yariv and Cross, 1979). More dissolves at high temperatures; at 50 °C and normal pH, solubility is 20 mg/L (Adamovic, 2005). At room temperature, the rate of quartz dissolution is extremely slow and controlled by the rate of bond breaking and hydration of silica at the surface of the mineral (Martini, 2000a) (see also the review by Dove and Rimstidt, 1994). In laboratory conditions at 25 °C, the rate of dissolution is 10^{-17} mol/cm^2/s (Bennett, 1991). But identifying the form of silica does not give an easy answer as to its solubility in weathering situations. Surface disruptions on quartz crystals can impart a solubility more akin to that of amorphous silica; clay minerals, soil particles and even quartz itself can sorb silica; amorphous silica can sorb cations such as iron and aluminium, organic anions and clay minerals; various salts can alter the rates and equilibrium values of silica solubility (Yariv and Cross, 1979).

In cold, neutral-pH groundwaters, silica concentration usually does not exceed 20–30 mg/L, but at higher temperatures, silica concentration increases and may reach thousands of milligrams per litre in some hot springs (Serezhinikov, 1989). Interestingly, while limestones are generally regarded as being much more soluble than silica, in *pure* water the most calcite ($CaCO_3$) that can go into solution is only about 13 mg/L at 16 °C and 15 mg/L at 25 °C (Jennings, 1985). This is only slightly higher than quartz. The greater prominence of limestone solution relies on the fact that most natural waters contain acids, notably carbonic acid from dissolved carbon dioxide, and this raises the solubility of calcite greatly, generally into the range of 250–350 mg/L at normal temperatures (Jennings, 1985). No common natural factor increases the solubility of silica so dramatically.

In brief, therefore, the form in which silica is present influences both the solubility and the dissolution processes. Silicate minerals weather more easily than quartz, and amorphous silica is more soluble than quartz. Nevertheless, silica dissolution is an important weathering process, not only in felspathic sandstones or sandstones with clay cements, but also in highly quartzose sandstones. Also, the degree to which silica in its various forms is naturally soluble in water is influenced by a range of factors, including pH, the presence of other reactive species, and, as already noted, temperature (Siever, 1962; Yariv and Cross, 1979).

The effects of pH

Correns (1941) suggested that the solubility of amorphous silica steadily increased above pH 5, but Alexander (1954, quoted in Siever, 1962, p. 128) considered the solubility of amorphous silica as monomeric silicic acid to be independent of pH from pH values 2 to about 9.5 (Fig. 5.2). As Fig. 5.2 shows, the solubility of silica is essentially stable between pH 3 and pH 8. In highly alkaline solute, the solubility rises exponentially. At the very acidic low end of the pH scale, the solubility of silica is believed also to rapidly increase, but unfortunately the solubility of silica in this region of low pH has received little attention (Serezhinikov, 1989; Brady and Walther, 1990). However, such extreme acidity is rare in natural environments (although in acid sulphate soils, the destruction of clays releases not only aluminium but also silica into soil water).

Laboratory studies have shown that at pH values above about 9.5, the solubility of silica increases very rapidly into the range of 600–1000 mg/L (Yariv and Cross, 1979), mainly as a result of ionization of H_4SiO_4 (equation 5.3) (Martini, 2000a). The value is actually pH 9.83, corresponding to the first dissociation constant of silicic acid, with quartz solubility reaching values of >20 mg/L at pH 10 and 25 °C (Adamovic, 2005). Laboratory and field studies in karstic

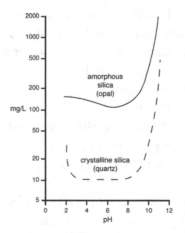

Fig. 5.2 The relationship of the solubility of silica to pH (after Krauskopf, 1956; Iller, 1979; Serezhinikov, 1989)

landscapes on sandstones and quartzites in Brazil (Wiegand *et al.*, 2004) revealed a strong correlation between pH value and quartz solubility. This study showed a strong solution maximum of SiO_2 in groundwater at around pH 5 (investigated range 3.5–7.5), but peak dissolved silica values were still less than 8 mg/L.

Inorganic salts and metal ions

A number of naturally occurring chemical species also affect the dynamics of the silica–water reaction system. This may lead to a marked divergence from total equilibrium solubility concentrations of silica, and also the rate of solution that might be expected from laboratory modelling. Siever (1962, p.135) concluded that 'the behaviour of natural materials in the geologic environment cannot be precisely judged by comparison with pure materials in the laboratory and that other chemical components may be affecting the equilibria'.

It is well known that a range of inorganic salts have a physical effect on sandstone, but some also influence chemical weathering. A. R. M. Young (1987) summarized the complex effects of salts on silica solubility:

- Solubility is decreased by increasing concentrations of most inorganic salts. Marshall and Warakomski (1980) demonstrated an increasing 'salting out' of silica by increasing concentration from 0–7 mol/L for a wide range of salts.
- Multivalent cations such as Al^{3+} were the most effective agents (Iller, 1979; Okamoto *et al.*, 1957), and the anion species was of little significance.
- Thus, the equilibrium solubility of silica in the presence of most salts at high concentrations will be very low, particularly in the absence of appreciable carbonate or phosphate to provide high pH values.
- However, the presence of sodium chloride has a strong accelerating effect on the rate of dissolution of silica. Increasing chloride or sulphate ion concentration accelerates the dissolution rate of crystalline silica (Yariv and Cross, 1979), whereas increasing pH (between 5 and 11) and increasing sodium chloride concentration accelerates the dissolution rate of amorphous silica (Kastner, 1981).

R. W. Young (1988) also argued that the intense dissolution of silica in the sandstones of the Bungle Bungle Range, Western Australia, may be enhanced by high chloride concentrations, and showed supporting SEM images of salt crystals on etched quartz surfaces (Fig. 5.3). This field evidence is supported by laboratory results that indicate either or both the solubility and the rate of solution of quartz can be increased by increased concentrations of Group 1A salts in the solute (Dove and Crerar, 1990; Bennett, 1991; Dove and Elston, 1992). In contrast to an earlier assertion by Siever (1962), von Damm *et al.* (1991) and Dove and Elston (1992) found that quartz solubility is increased in seawater (about 0.5M NaCl) relative to distilled water.

Chemical weathering

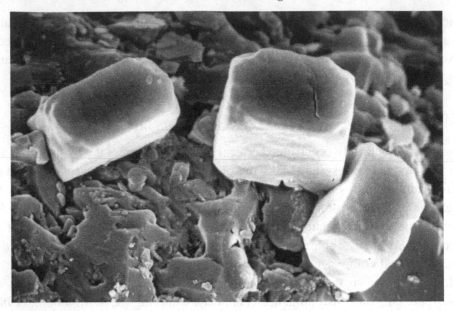

Fig. 5.3 SEM image of salt crystals rest on etched quartz grains (× 1200), Bungle Bungle Range, Australia

These findings seem to be in direct contrast to the assertions of other authors (e.g. Goudie, 1974) who insist weathering of sandstones in highly saline environments is dominantly a product of mechanical breakdown by salt crystal expansion. What is evident, though, are extremely high levels of quartz etching at these locations (see also McGreevy, 1985) and these solutional textures appear to be related to high sodium chloride concentrations in the manner proposed by A. R. M. Young (1987) and von Damm *et al.* (1991).

Some metal ions may also influence silica solubility. Yariv and Cross (1979) suggested that the solubility of monosilicic acid is not affected by the presence of monovalent metallic cations, but that multivalent metallic cations do change silicic acid solubility. Reardon (1979) also noted this fact and suggested that iron–silicate complexing, for example the formation of FeH_2SiO_4, may be significant in the mobilization of silica in acid waters. Morris and Fletcher (1987) tested the hypothesis that quartz solubility was increased markedly by quartz/ferrous iron reactions, concluding that a rapid release of silica may occur in some ferrous iron solutions under oxidizing conditions, and that the potential solubility of quartz may be increased by a factor of 10 over that of amorphous silica. Serezhnikov (1989) also found increased solubility of quartz in the presence of iron. Tripathi and Romani (2003) reported that small amounts of pyrite (FeS_2) promotes weathering in quartzite near Delhi, India. They postulated a coupled mechanism in which the dissolution of pyrite produces a sulphate-bearing acidic solution and ferrous iron, which react respectively with aluminosilicate minerals and quartz. This increases the porosity of the Delhi quartzite, promoting seepage, and making the rock more friable.

Complexing by other multivalent metal ions such as Mn and Al – the oxide phases of which are poorly soluble at moderate pH but more soluble at low pH – may also be important in complexing and mobilizing silica in acidic waters (Reardon, 1979). For aluminium, widely differing conclusions have been reached. Bennett *et al.* (1988) found no correlation between aluminium in solution and quartz dissolution, but Beckwith and Reeve (1969) and Mullis (1991) reported that low concentrations of Al^{3+} may strongly inhibit quartz dissolution. McFarlane and Twidale (1987) argued that the presence of Al^{3+} ions in solutions of silica will lead to immediate co-precipitation.

Organic acids

For over 115 years (Julien, 1879; also quoted in Bennett *et al.*, 1988), organic compounds have been thought to enhance the solubility of silica. Descriptions of anomalous silica movement in geological environments rich in dissolved organic compounds are commonly found, but a model of silica mobility incorporating both the organic and inorganic reaction mechanisms has been slow to appear in the geologic literature (Bennett *et al.*, 1988).

Siever (1962) suggested that some organic compounds may actually lower the solubility of amorphous silica by coating the reactive surface; but, on the other hand, certain organic compounds may also form organic–silica complexes which increase the solubility. Huang and Keller (1970) and Jackson *et al.* (1978) found that the solubility of many silica minerals was higher in the presence of organic acid anions than in water alone. The solubility of Al, Fe, Ca and Mg were at the same time also boosted. Yariv and Cross (1979) also found that the solubility of amorphous and crystalline silica increased slightly in the presence of certain organic acids. Obviously the organic acids attack not only the quartz grains and overgrowths but also other cement or matrix materials. A. R. M. Young (1987) suggested that cavernous weathering in the Hawkesbury Sandstone near Sydney was due in part to percolating water rich in organic acids taking iron oxide cement into solution.

In a study of oil-contaminated shallow groundwater, Bennett *et al.* (1988) showed that dissolved silica concentrations correlate with concentrations of dissolved organic carbon. Furthermore, they observed that the reactivity of quartz with several organic acids could be defined by the reaction series: citrate > oxalate > salicylate > acetate (see also Bennett, 1991). This series parallels the one found by Huang and Keller (1970) for organic acids that form complexes with aluminium and accelerate the dissolution of alumino-silicates at acidic pHs.

Interestingly, it seems that only the multi-protic acids (oxalate and citrate) or the multi-functional acids (salicylic acid) accelerate quartz dissolution, whereas acetic acid, a mono-functional mono-protic acid does not (Bennett *et al.*, 1988). In the presence of citrate, the quartz dissolution rate increases up to $10^{-15.5}$ mol/cm^2/s (Bennett, 1991). Ghosh (1991) claimed that multi-functional organic acids may accelerate quartz dissolution by decreasing the activation energy.

Also, several polar organic anions increase both the rate of quartz dissolution and the apparent equilibrium solubility in aqueous systems by chelating silicic acid in solution, thereby decreasing the activity of the monomeric silicic acid, and resulting in an increase in solubility. Where dissolved organic carbon is high, temperatures are low, and the pH is buffered to near-neutral conditions, organic–silica interactions may be an important process (Bennett, 1991). Field studies in the Brazilian quartz sandstone karst (Wiegand *et al.*, 2004) did not support a postulated increase in quartz solubility due to formation of silica–organic compounds, but laboratory work did show accelerated dissolution when organic concentrations were high. Martini (2000a), noting that waters off most quartzites have low pH because of organic acids, concluded that the rate of dissolution may be increased, but organic complexing with silica is only effective at high pH.

Bioweathering

That some plants and fungi seem to directly assist the chemical and physical breakdown of silicate rocks has long been known, but interest in the nature of biodeterioration of stone, and in the possible roles of biological agents in boosting the breakdown of natural and man-made stone surfaces, has increased in recent years (Papida *et al.*, 2000; Stretch and Viles, 2002; Burford *et al.*, 2003; Turkington and Paradise, 2005; Hoffman and Darienko, 2005). Several papers have provided comprehensive reviews of the processes involved in biodeterioration (e.g. Kumar and Kumar, 1999; Warscheid and Braams, 2000; Hoffland *et al.*, 2004).

Lichens, algae and microbial activity

That lichens and other biological growths can be important in sandstone weathering has been known for more than 85 years, since the investigations of Fry (1922, 1924). Although Viles and Pentecost (1994) noted that we are still far from a complete understanding of the interaction of lichens, weathering and landform production, there is general agreement that low-order biological organisms generally degrade sandstone surfaces (Wessels and Schoeman, 1988; Turkington and Paradise, 2005). Henderson and Duff (1963), for example, noted that some fungi and bacteria produce solutions high in organic acids. The cumulative effects

Table 5.1 *Bioweathering and landforms at different scales*

Scale	Landforms	Processes	Controlling factors
μm	Fungal boreholes and etching under lichen thalli	Chemical attack by lichen acids, carbon dioxide	Crystallography; Lichen species; Micro-environment
mm	Fruiting body pits Grooves Case hardening and altered zones	Chemical attack by lichens + grazing by snails, etc. Chemical attack by lichens + inorganic chemical weathering	Microscale rock variability; Aspect; Micro-climate; Lichen community dynamics
cm	Weathering basins	Biochemical attack + various inorganic weathering processes	Jointing and bedding planes; Slope angle; Lichen and algae zonation
> m	Ruiniform relief	Weathering and erosion	Case hardening; Long-term etching of quartz; Rock characteristics; Climate and climate change

Source: Viles and Penticost (1994).

of such small-scale reactions over long periods of time could contribute significantly to the weathering of minerals.

Viles and Penticost (1994) provided an outline of bioweathering processes on landforms across a range of scales (Table 5.1).

Bjelland and Thorseth (2002) conducted a detailed study of lichen weathering on sandstones of western Norway, finding both chemical and physical differences between different lichen species. Whilst alive, the thallus and endolithic fungal hyphae may bind partly fragmented rock surfaces, protecting them from abrasion and erosion. When lichens die, the loose material will be eroded and a new sandstone surface exposed. Paradise (1997) noted the unequal weathering of sandstone under, compared to beside, lichens in Arizona. Beneath lichens where rhyzine density is greatest, rhyzine penetration physically disaggregated the sandstone matrix, but little chemical weathering was apparent. But chemical weathering increased from the cortex to the thallus fringe with the transport (by dew and precipitation) of lichen acids from the top of the lichen to the growing margin. He found that lichens may change the host rock both by producing acids and by the physical penetration of rhyzines. As the thalli enlarge (i.e. the lichen spreads), rhyzine penetration accelerates the disintegration of the previously chemically weathered rock (Fig. 5.4).

Fig. 5.4 Pale crusts of lichen and thin bedding planes defined by Liesegang rings
create patterns on a Hawkesbury Sandstone surface darkened by finely dispersed
lichens, Sydney Basin, Australia

In Morocco, Duane (2006) reported on the breakdown of sandstones by
lichens. Mechanical and chemical breakdown are both apparent, with quartz
grains weathering by the production of peeling structures and micro-brecciation
features. Dissolution on quartz crystal surfaces appears to be a surface reaction-
controlled process mediated by microbial microfilaments and nanofilaments.
Rock breakdown is a complex process, probably beginning at the nanoscale with
penetration by the fungi of sites on crystal faces. In the High Atlas Mountains of
Morocco, Robinson and Williams (2000) found surface weathering features
associated with the growth of a species of *Aspicilia* lichen. Weathering is most
active at the margins of the thalli; and as the lichens grow, they incise into the
sandstone forming circular or crescentic depressions up to 10 mm deep. The
physical undercutting or sapping of individual quartz grains at the depression
microscarp is the most active process. It may also be assisted by the dissolution of
iron and other cementing material by lichen acids produced by the top of the
thallus and transported to the fringe of the growing thallus in a manner similar to
that identified by Paradise (1997). Rates of rock loss were estimated at between
0.5–4.6 kg per lichen per century, equivalent to a downwearing rate of 60–220
μm (0.06–0.2 mm) per year. This is of the same order of magnitude as weathering

by lichens measured on Clarens Sandstone in South Africa (Wessels and Schoemann, 1988; see also Cooks and Pretorius, 1987; Cooks and Otto, 1990) and the Beacon Sandstone formation of Victoria Land, Antarctica (Hale, 1987).

However, lichens do not always accelerate surface weathering. Coates (1989) found that lichens in the Sydney region have little effect on the disintegration of the Hawkesbury Sandstone, and several authors have suggested a protective rather than a destructive role for epilithic lichens on some rocks. Kurtz and Netoff (2001) found sandstone surfaces in south-central Utah were protected by surface micro-organisms; and at Petra in Jordan, Paradise (2003) noted that lichen growth (*Lechanora* species) on north-facing surfaces actually causes recession rates to fall off dramatically. He argued that while lichens accelerate subsurface weathering, they may also – at Petra at least – decrease surface erosion (apparently by acting as a protective covering). Viles and Penticost (1994) drew attention to R. W. Young's (1987) observations of algal layers protecting the friable sandstone surface in the Bungle Bungle Range of the Kimberley. They believed that a similar process may also be occurring in the Cedarburg Mountains in South Africa, where epilithic lichen cover etches quartz grains, 'at the same time (until their death or removal by grazers) providing an overall protective cover a few hundreds of micrometres thick' (Viles and Penticost, 1994, p. 113). We note that the 'algae' in the Bungle Bungles have been identified as cyanobacteria (Hoatson *et al.*, 1997).

In the hyper-arid Al-Quwayra area of south Jordan, subsurface algae appeared to strengthen areas of case-hardened friable white Cambrian/Ordovician sandstone, reducing rates of erosion significantly in comparison to areas which have no algal growths (Goudie *et al.*, 2002). Crypto-endolithic biofilms containing cyanobacteria and fungi are often associated with a range of types of iron, manganese and calcium surface case hardening (about <1 mm thick) on Cambrian and Ordovician sandstones (Viles and Goudie, 2004). While micro-organic biofilms may aid the formation of the case hardening through enhancing cementation, they are by no means a necessary component of all the hardened layers in the region. A number of researchers (e.g. Warscheid and Braams, 2000) have also argued that microbial contamination contributes significantly to the acceleration of weathering processes, but this is not uniform. Mottershead *et al.* (2003) found that abundant micro-organism growth protects rock, but patchy growth leads to enhanced weathering. Turkington and Paradise (2005, p. 242) noted several roles for biofilms:

- exacerbating the damage caused by salts;
- promoting salt and frost attack by increasing pore volume, moisture content, mineral alteration, and precipitation of oxalates or sulphates;
- or, conversely, by reducing effective porosity/permeability and thus inhibiting salt weathering and pollutant ingress to the rock;
- altering the thermal response and wetting of the rock.

Silica transport in streams and groundwater

Surface and groundwater analyses demonstrate the almost universally low solubility of quartz in natural environments. Quartzite yields waters which are usually very pure, with little dissolved load, and silica as the dominant solute. For example, Martini (2000a) noted that silica concentrations in groundwater from the quartzites of the Table Mountain Formation in South Africa ranged from 1–30 mg/L for 17 samples, averaging 6.3 mg/L, and were thus undersaturated with respect to quartz; the single high 30 mg/L value was possibly due to impurities such as mica or feldspar, or to concentration by evaporation. In tropical high-rainfall Venezuela, waters draining from the Roraima quartzites also typically display very low concentrations – <1 mg/L in table-mountain waters (Chalcraft and Pye, 1984), 0.18–0.52 mg/L in runoff and in underground fractures to 80 m depth, and 0.92–1.30 mg/L in deep (300–350 m) fractures (Mecchia and Piccini, 1999). Mecchia and Piccini suggested that 15% of silica came from surface water solution and 85% from underground water solution. Even in the streams in the valleys below the table mountains, levels are low. Chalcraft and Pye (1984) found a range of 5–7 mg/L; 2.56–4.27 mg/L has been recorded in the Caroni Basin streams draining the Roraima quartzites and associated mafic intrusions.

On a broad scale, temperature seems to be an important factor influencing silica solubility, for as in many reactions, the solubility of silica does increase with temperature (Siever, 1962). Livingstone (1963) suggested a global relationship – average concentrations are lowest for arctic rivers (5 ppm), higher for temperate rivers (10 ppm), and highest for arid and tropical humid rivers (20 ppm). Similarly, Meybeck (1987) found an average dissolved silica concentration of only 3 mg/L for arctic streams, in contrast to 8.4 mg/L (0.14 mmol/L) for temperate streams, and 13.2 mg/L (0.22 mmol/L) for tropical streams. Meybeck also found that within a temperate area, silica concentrations from basins of several lithologies correlated negatively with altitude and hence temperature.

Meybeck concluded that 74% of the silica load to oceans comes from the tropical region, a fact that depends both on rate of solution – which is temperature-dependent – and on discharge. This claim is, however, at odds with Douglas (1969, 1978), Davis (1964) and Thomas (1974). They argued that there is no conclusive evidence for high silica concentrations in tropical rivers; and therefore any suggestion that a high rate of silica removal takes place in the tropics must rest upon evidence of higher stream discharges. Douglas (1969) established a significant relationship between total silica load and runoff, and it would thus appear that high rates of chemical denudation as indexed by silica removal depend largely on runoff (Thomas, 1974). And Douglas (1969, 1978) considered that the more rapid decay of organic matter, and a consequent increase in the interaction of silicates and organic acids in the humid tropics could offset any direct effect of temperature. Thus the volume of water, rather than temperature, appears to be the critical factor in the removal of silica from siliceous rocks.

Flushing rate

The bulk removal of silica is therefore dependent not only on its solubility, but also on the rate at which water moves through the rock. Douglas (1978, p. 230) commented that 'the importance for the rate of solution of relatively rapidly moving water has been demonstrated in limestone terrains . . . but it is equally significant in silicate areas. . . . the rate of loss of ions from silicate minerals to waters is controlled by the speed at which dissolved ions are carried away from the surface of the mineral'. Rimstidt and Barnes (1980) also emphasized the importance of the flushing rate; and Doerr (1999) also noted that a high precipitation rate and water throughput is required to counter the low total solubility of silica.

It is obvious that, for a given solubility regime, higher rates of flushing will increase silica loss. Velbel (1985) suggested that the flushing rate also influences the mineralogy of the weathering mantle. If the flushing rate is high through soil or saprolite in which feldspars are being weathered, silica concentrations remain low but aluminium hydroxide may precipitate out; when the flushing rate is lower, silica and alumina levels may be high enough for kaolinite to form. Thus with high rates, silica remains mobile and may enter stream flow, whereas with low rates, it may – temporarily at least – be 'fixed' by neo-formation of clay. The importance of the flushing rate was emphasized by Thiry *et al.* (1988) and Thiry (2005, 2006) in their studies of the Oligocene Fontainebleau Sand of the Paris Basin. In the vadose zone above the groundwater table where flushing occurs, silica is leached out and the sand grains are corroded. In the phreatic zone below the water table, the sands are in constant contact with the porewater, and silica reaches supersaturation. Rather than being leached out, it is precipitated as overgrowths on quartz grains at a geologically rapid rate. They estimated that a quartzite (silcrete) lens 2–8 m thick develops in about 30,000 years.

However, an increase in flushing rate will not necessarily give the same rate of increase in dissolved silica. Berner (1978) emphasized that flushing accelerates the dissolution of minerals only up to a limiting rate beyond which additional through-flow of water has virtually no effect. Infinite silica cannot be provided because dissolution is controlled by mineral reactivity. Mainguet (1972) recorded the results of experiments using feldspathic sandstone (75% quartz, 24% feldspar, 1% mica) to look at solutional loss over a 3-year period. Silica loss averaged 4.25 mg/L for experiments carried out at 30 °C and 7.15 mg/L for those carried out at 50 °C; the loss of Fe and Al was insignificant at both temperatures. Silica loads of natural streams were found to be of the same order; but interestingly in African streams (in southern RCA), the silica concentrations were highest at the end of the dry season and declined by a few milligrams per litre during the wet season. Mainguet considered that this was due to the waters issuing at the end of the dry season having had the longest residence time in the rock.

Given the different solubilities of silica from varying mineralogies, it is surprising that Meybeck (1987) suggested that silica loads in streams on a global scale do not vary greatly with catchment lithology. His data suggested that percentages of total silica load from sandstones, granites, gneisses/mica schists and shales almost equal the percentages of outcrop area of these rock types. Certainly, at catchment level, the picture is not so simple. For example, in the Amazon Basin, streams draining the steep slopes on silicate rocks in the Andes have high suspended loads and higher silica loads than streams draining the more quartzose lithologies of planated lowlands. The 'whitewater' streams average 4.1 mg/L (0.07 mmol/L) silica, whereas the 'clearwater' tributaries further downstream average only 1.9 mg/L (0.03 mmol/L). When the clearwater streams flow through swampy lowlands, the presence of high organic loads discolours the water and also reduces the silica load to an average 0.6 mg/L (0.01 mmol/L) in 'blackwater' streams (Brinkmann, 1986). It is not only lithology and organic load which are important. The thin soils and broken slopes of the Andean streams increase the potential for weathering of the silicate rocks in this part of the catchment, and thence for higher silica loads in whitewater streams; the thick soils, deeply weathered surfaces and gentle slopes in the lowland catchments contribute to low dissolved loads there (Stallard, 1985). Temperature, lithology, weathering maturity, vegetation and topography all influence the mobility of silica in the environment.

The locus of chemical attack

How, then, is quartz sandstone affected by solution? At the microscopic level, many authors have noted direct evidence for the chemical attack and etching of the surface of quartz grains and overgrowths under both field and laboratory conditions. Crook (1968) argued that embayments and smoothed surfaces on detrital quartz grains cannot be explained by physical weathering, but only by solutional attack. Burley and Kantorowicz (1986) described quartz grain surface features including pits, notches and embayments produced by solutional attack, with a tendency for corrosion to be more intense on the surfaces with high free surface energy (see also Hurst and Bjorkum, 1986). Free-energy sites are common around grain peripheries, along fractures and between crystal boundaries in rock fragments. They proposed two mechanisms of quartz corrosion:

- Transport-controlled dissolution is controlled by the rate of transport of ions to and away from the reaction surface. It is characterized by rapid, non-specific corrosion and is typical of strongly concentrated solutions or highly soluble minerals. It gives rise to intense etching and corrosion of all available sites.

- Surface reaction-controlled dissolution, by contrast, is controlled by the reaction rate at the solid–fluid interface. It is generally slow and more specific, being typical of slow dissolution of relatively insoluble minerals in solutions of low chemical reactivity. Surface reaction-controlled dissolution thus tends to produce distinct crystallographically controlled features such as well-defined notches.

R. W. Young (1988) found both surface reaction and transport-controlled dissolution features in a SEM investigation of the regional extent and intensity of quartz sandstone etching in the East Kimberley of Western Australia (see Figs 1.6 and 1.7). He also noted that Hurst and Bjorkum (1986) had challenged the ideas of Burley and Kantorowicz (1986), arguing that quartz dissolution rates are too low for transport-controlled etching. Hurst and Bjorkum emphasized that etching will concentrate at sites with the highest free-energy, and quartz overgrowth lowers the surface free-energy of a detrital grain. According to Hurst and Bjorkum (1986), dissolution will therefore be most rapid at the greatest concentration of detrital grain surfaces, face-corners and edges of overgrowths. Brady and Walther (1990) and Withe and Peterson (1990) supported the proposition that silicate dissolution occurs preferentially at high-energy surface sites such as defects, and is controlled by the density of such defects.

White *et al.* (1966) argued that in the hot and extremely wet conditions found in tropical Venezuela (areas which receive up to 7500 mm annual average precipitation), quartz within the Roraima quartzites was hydrated to much more soluble opal, then removed in solution. However, Martini (1979) stated that the transformation of quartz to opal is thermodynamically not possible at surface conditions. Others have proposed that the quartzite had been hydrothermally altered along fissures (Szczerban *et al.*, 1977) – a possibility locally but unlikely as an explanation at a regional scale (Martini, 2000a). Chalcraft and Pye (1984) also rejected a hydration mechanism, and showed direct solution of quartz, without the intermediate hydration to opal occurring. They also showed that – while there is preferential solution along joints, beds and lithological contacts – cracks at all scales are foci for water flow and thus pathways for solution. Thin sections showed widening of grain-to-grain contacts along with etching and corrosion of both quartz grains and cements, eventually leading to a removal of cement and the freeing of individual detrital grains; SEM examination demonstrated intense microscopic pitting of quartz grains and overgrowths (Chalcraft and Pye, 1984). Ghosh (1991), also working in the Roraima, found that the unweathered rocks show abundant welding of grains by a pervasive syntaxial quartz cement and by sutured grain-to-grain contacts. In contrast, the weathered samples display an excellent network of lamellar porosity formed by dissolution of quartz cement along mainly adjacent overgrowth boundaries. This process is similar to that recognized by petroleum geologists in the formation of secondary sandstone porosity by quartz dissolution (Pye and Frinsley, 1985; Burley and

Kantorowicz, 1986; Hurst and Bjorkum, 1986; Shanmugam and Higgins, 1988), except that it is occurring at near-surface conditions.

There is also clear evidence for solutional etching of quartz in sandstone areas outside the humid tropics – in the cool temperate Millstone Grit, England (Wilson, 1979), the tropical semi-arid Kimberley region of Western Australia (R. W. Young, 1986, 1988), and the warm temperate Sydney Basin in south-eastern Australia (Wray, 1997a, 1997c). In the quartz sandstones of the Sydney Basin, Wray identified the two types of microscopic solutional attack which had been noted in quartzites and quartz sandstones from tropical regions (Chalcraft and Pye, 1984; R. W. Young, 1986, 1988):

- The small, often 'v'-shaped, pits on grain and overgrowth surfaces that show strong crystallographic control (denoting surface reaction controlled dissolution).
- The larger, irregular, embayments or depressions with no crystallographic control that often penetrate through quartz overgrowths into the grains below (denoting transport-controlled dissolution).

Wray (1997a, 1997b) noted that not all sites within the rocks are attacked equally. Detrital quartz grains, grain-to-overgrowth boundaries, overgrowth-to-overgrowth contacts and other similar discontinuities or defects are generally far more corroded than most overgrowth faces (Fig. 5.5). And etching of quartz overgrowths is widespread and most intense on the rhombohedral faces and edges,

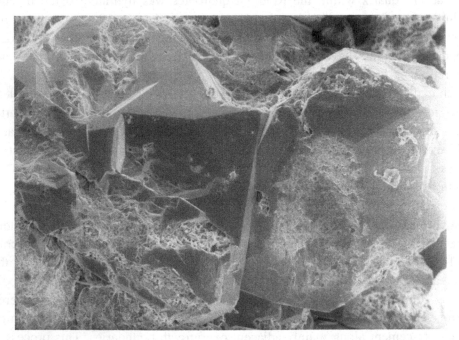

Fig. 5.5 SEM image of intense etching of grains and quartz overgrowths, Nowra Sandstone, Sydney Basin

especially in the most weathered sandstones, whilst there is less attack on the overgrowth prism faces. The apparent crystallographic etching can be explained in terms of variable free-energy, as the overgrowths have a lower free surface energy than detrital grain or boundary surfaces (Hurst and Bjorkum, 1986).

Wray also argued that the variability in the intensity and type of quartz etching in the Sydney region may be linked to the degree of primary porosity, both in the amount of interlocking quartz overgrowth, and the proportion of void-filling authigenic clays. As in the sandstones of the Kimberley region (R. W. Young, 1988), it is the primary porosity that has the most influence on the penetration of corrosive solutions throughout a sandstone. Sandstones with little interconnected void space provide few pathways for water penetration and are often only mildly weathered and mostly close to the surface; whilst rocks with a high degree of permeability are usually deeply weathered throughout the sandstone, and very highly etched at the microscopic scale.

Speleothems

Much of the silica dissolved from sandstone is flushed out of the rock, and into streams or groundwater. However, some is re-precipitated, or incorporated into new minerals such as by neo-formation of clays in soils. Sandy stream deposits may also be re-cemented by silica derived from other parts of the sedimentary mass, or at least from nearby, to form silcretes.

Far less well known are the speleothems that characterize many quartzose sandstone landforms. Speleothems are common in most carbonate caves, where they are typically dominated by calcite. But in areas of siliceous rocks, and in the absence of significant carbonate cement, speleothems of silica are generally smaller but still quite common. They assume a variety of forms which mimic carbonate speleothems – including anthodites, blisters, boxwork, coralloids, crusts, flowstone, stalagmites, stalactites, columns and even helictites (Hill and Forti, 1986). Silica speleothems have been found in granite and basalt caves (Webb and Finlayson, 1984; Willems *et al.*, 2002). They have also been described within numerous sandstone and quartzite caves worldwide, and provide clear evidence of their solutional genesis (Wray, 1997b). The most numerous and detailed investigations to date have been from the caves and caverns of the northern South American quartzites, particularly Roraima, where this speleothem type is quite common.

Silica speleothems

The first reports of flowstones and stalactites in the Roraima quartzites were by White *et al.* (1966). They argued that thin coatings of opal flowstone and small stalactites on the wall of a small fissure provided evidence for the solution and later re-deposition of silica. Little subsequent investigation of these silica speleothems was undertaken for almost a decade until the review by Urbani and Szczerban (1974). In the caves of the Sarisariñama Plateau, numerous speleothems were identified, including opal stalactites up to 10 cm long, flowstones and different types of crusts and coral-like shapes (Zawidzki *et al.*, 1976). Stalactites, stalagmites and coralloid speleothems from Cerro Autana Cave showed concentric growth banding composed of opal, length-fast chalcedony and calcite. These deposits were thought to originate from direct precipitation of the opal and calcite from alkaline waters at ambient temperature, with the chalcedony representing subsequent recrystallization of the opal. Some components were believed to have originated from the breakdown of feldspars in arkose horizons within the quartzites (Urbani, 1976). Urbani (1977) also reported siliceous stalactites in caves near Kumerau Fallas, and in virtually all caves in the Roraima Formation.

Flowstone and stalactites from Mount Roraima analysed by Chalcraft and Pye (1984) were found to differ from those previously described. The flowstone had a porous texture and consisted of detrital sand grains cemented by cristobalite, tridymite and authigenic quartz (Chalcraft and Pye, 1984) instead of the opal-A matrix found by most of the earlier researchers. Urbani (1990, 1996) reviewed published literature on speleothem and other cave minerals found in the Precambrian non-carbonate rocks of the Venezuelan Guyanan Shield. He noted that speleothems from several mineral groups have been found in Venezuela, including carbonate (calcite), silica species (allophane, chalcedony and opal-A), nitrates (nitrammite and sveite), oxides/hydroxides (goethite and lithiophorite), phosphate (evansite) and sulphates (aluminite, epsomite, gypsum and hexahydrite) (Urbani, 1996).

Briceño and Schubert (1990) also reported small amorphous silica stalactites within the Roraima region's cavernous orthoquartzites. Carreño and Urbani (2004) described amorphous opal-A stalactites from the Roraima Sur cave system. There, microclimates affect air flow patterns, condensation and capillary action, and appear to cause zonation between the entrance and deeper parts of the cave, modifying speleothem formation and establishing varying patterns of dimension, concentration, spatial distribution and rotation angles due to wind. Wiegand *et al.* (2004) report speleothems – including coralloids, crusts, flowstones, stalactites and stalagmites composed of opal and the rare mineral, varescite, within caves in quartzites of several areas in eastern Brazil.

There have been fewer studies of silica speleothems in sandstone caves outside South America. Martini (1979, 1982) described small stalactites composed of limonite from the Black Reef Quartzite in South Africa. Opal 'popcorn' was also found, as well as a dark, partly organic, flowstone. Porter (1979) noted opaline stalactites and coral on sandstones of the Lee Formation in Virginia, USA. In Europe, Urban *et al.* (2006) described speleothems from the flysch sandstone caves of the Polish Beskidy Mts. Most speleothems came from caves on Mt Kilanowska and neighbouring Mt Cergowa. Large and abundant calcite speleothems of Holocene age are found in the region, but also a number of non-calcite speleothems from gravitational fissure caves in sandstones high in carbonate. In the Roznów foothills, another cave in quartz sandstone exhibited thin incrustations and botryoidal forms composed of kaolinite, quartz and amorphous material with admixture of chlorites. In other Polish caves, light brown stalactites 10 cm long are formed externally of thin laminates of an unknown amorphous material with silica in various stages of crystallization (opal–quartz).

Reports from Australia have indicated the widespread occurrence of silica-based speleothems. While Jennings (1979) reported small stalactites on sandstones in northern Australia, most published accounts from that region have concentrated on flowstones. These silica 'skins' have been studied in detail because of their association with aboriginal rock art. In Kakadu National Park, Watchman's (1992, 2007) detailed analyses of numerous silica skins, or flowstone, showed silica associated with a wide range of other minerals, including various compounds of calcium, silica, sulphur, aluminium, magnesium and potassium. Silica speleothems have been reported from numerous sandstone overhangs and caves in southeastern Australia, especially near Sydney. A. R. M. Young (1987) noted that small (up to 10 mm long) silica stalactites were common in both coastal and inland sites; and Young and Young (1992) emphasized that these small stalactites on the roofs of caverns clearly demonstrated the movement of silica in solution through the sandstones of the Sydney Basin. Generally the stalactites form under overhangs which have roughly planar roofs, due to extension back along bedding planes, and they occur across the roof and not along lines of obvious seepage or in linear arrangements. Hence they indicate seepage through the rock mass, rather than seepage along fractures or joints (Fig. 5.6).

Speleothems are widespread in the Sydney Basin sandstones (Wray, 1997c, 1999). Silica is depositing from solution in groundwater as a variety of small speleothems; most commonly conical and coralloid stalactites, but flowstones and rare stalagmites are also forming (Fig. 5.7). These speleothems are all composed of banded silica. Calcite is absent, and opal-A, chalcedony and kaolinitic clays are the only dominant minerals. Evaporation of water seeping through the rock initially deposits dissolved silica as tiny flecks of amorphous opal-A (Fig. 5.8). Successive layers of opal-A result in the growth of stalactites and stalagmites. Over time, opal-A can convert to cryptocrystalline chalcedony (Morse and Casey, 1988), but the rate of conversion is not known (Wray, 1999) (Fig. 5.9). These

Fig. 5.6 Under an overhang in Hawkesbury Sandstone, small silica stalactites hang below a boulder. Pale flowstone covers the boulder, which is about 0.5 m across

Fig. 5.7 A cluster of coralline stalactites from the Blue Mts, western Sydney Basin

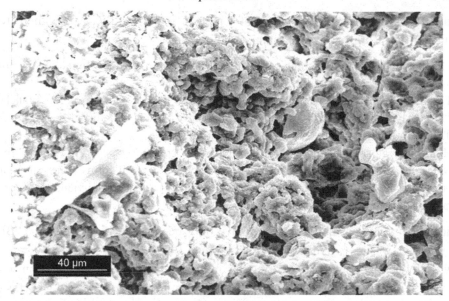

Fig. 5.8 SEM image of the surface of a conical stalactite. The surface is composed of myriad irregularly packed flecks of amorphous opal-A

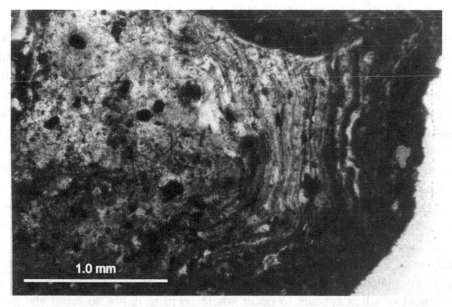

Fig. 5.9 Optical thin-section micrograph of a silica stalagmite. Dark layers are amorphous opal-A, and transparent layers are crystalline chalcedony. Field of view 3.95 mm

findings are similar to the findings of Urbani (1976) for silica speleothem formation in the Roraima. Three main stalactite forms occur:

- The first are irregular branching coralline stalactites from less than 1 mm to over 75 mm in length, with branches from 0.25 mm to over 12 mm in diameter.
- The second are conical, tapering from the base to the tip with no or few branches. These have no central drip-water hole, and have typical lengths of 1–50 mm and diameters from 0.5–5 mm.
- Highly irregular and highly porous, bulbous, coralline 'popcorn' is the most widespread stalactite form and is found at many locations. It forms either singularly or in bunches, often in association with other silica stalactites. Individual 'popcorns' range from around 5–50 mm in diameter.

Silica stalagmites are quite rare, and have only been noted by Wray in the Blue Mountains, west of Sydney (Wray, 1997c, 1999). They take the form of low, bulbous, hemispherical or irregular mounds, 5–50 mm in height. Some are surrounded by silica flowstone. A rare few have branching or coralloid growths on them. Wray (1997a, 1997b, 1997c, 1999) found that grey and white silica flowstones, 2–8 mm thick, occur widely in sandstones of the Sydney region, including the Nowra Sandstone which had also been reported by Watchman (1992).

Iron-rich speleothems

Lassak (1970) observed small stalactites of laminar limonite alternating with opal on quartz sandstones near Sydney. Young and Young (1992) also noted that stalactites in the Sydney Basin sandstones are more commonly iron-rich than siliceous; and where there is seepage emerging from joints or bedding planes, there is usually a gelatinous orange–red deposit which often drapes as a tufa over roughly horizontal surfaces (such as near the bases of waterfalls), hangs as smooth conical stalactites below the roofs of caverns or over the lips of undercut outcrops, or flows as narrow fans down the faces of cliffs (Fig. 5.10). The iron content of these forms is often high; one analysis yielded 172.8 g/kg amorphous iron. The deposits show flow structures and can be distinctly banded, but are iron-rich with only minor silica or clays (Fig. 5.11). Young and Young (1992) suggested that both iron and silica move in solution through the sandstones, but are precipitated in different environments. The movement of iron is more obvious and is associated with strong seepage along joints or partings in the rock; the movement of silica appears to occur where seepage is not so rapid and to be largely independent of partings or joints. Since the small silica stalactites on cavern roofs have no iron associated with them and the iron-rich precipitates have low silica contents, the geochemical conditions under which iron and silica move

Fig. 5.10 Smoothly rippled orange–red iron oxide tufa 0.5 m high and stalactites 0.3 m long have formed where seepage emerges along a joint in a cliff in the Budawang Range, southern Sydney Basin. (Photo: G. Chapman)

into solution and are then re-precipitated appear to be different. Both processes are important in the weathering of the Sydney Basin sandstones, and possibly in the development of cavernous weathering. It should be noted that Liesegang rings, the rhythmic iron oxide banding also common in many sandstones, probably formed during lithification, rather than being a weathering phenomenon.

Willems *et al.* (1998) regarded ferrous cave speleothems in Niger as evidence of the ferro-siliceous interactions leading to karst development there. Iron-rich speleothems have also been found in South America. Laffer (1973) had noted the formation of iron oxide speleothems within a cave in the 'Iron Formations' rocks of the Imataca Formation near the Tocoma River, and further silica and iron-based stalactites, stalagmites and flowstones were later discovered (Zawidzki

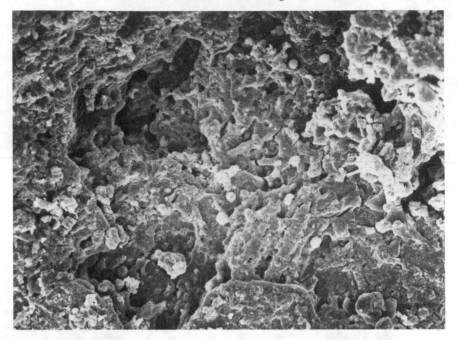

Fig. 5.11 SEM image showing the porous structure of iron oxide tufa

et al., 1976) in the caves and shafts of the Sarisariñama Plateau. Large goethite stalagmites several metres high were also discovered together with stalactites of more unusual minerals, including lithiophorite. An excellent example from the Sima de la Lluvia is illustrated in Figure 4, p. 35 of Zawidzki *et al.* (1976). Thus, the formation of both iron-rich and siliceous speleothems testifies to the movement of both iron and silica in solution through masses of sandstone, and to the importance of such movement in the development of karst in sandstone landscapes.

Speleothems and micro-organisms

There are interesting associations between micro-organisms and silica speleothems. Willems *et al.* (1998) found that bacteria may have assisted the formation of a deep phreatic sandstone karst in Niger. After the karst began to form, it seems that organic dust and bacteria depositions were important in secondary speleothem deposition processes. It is likely that micro-organisms may also play some role in the formation of Australian silica speleothems (Wray, 1999). Recent studies (Šmída *et al.*, 2005; Aubrecht *et al.*, 2008) show a range of about a dozen highly unusual opal-A and chalcedony speleothems from

Cueva Charles Brewer on the Chimantá Plateau, Venezuela. These speleothems are in fact fine-laminated columnar stromatolites, formed by silicified cyano-bacteria, surrounded by an outer, porous peloidal zone that was most likely formed by coccal bacteria. This unusual formative mechanism has prompted the use of the terms 'biospeleothems' and 'microbialites'.

Conclusion

In this chapter, we have dealt with the chemical processes that are involved in the weathering of sandstones, particularly quartzose sandstones. The solution reactions are complex and variable, and influenced by the range of pH, organic acid concentrations, presence of various anions and cations, and temperature. Speleothem development demonstrates the importance of the solution and movement of both silicon and iron in waters seeping through sandstone. The intensity of erosion is perhaps most affected by rainfall and thence the volume of water flowing through a landscape. The process of arenization – dissolution, granular disintegration and erosion – is critical to the development of many features of sandstone landscapes. And it is to the specific landforms formed by processes driven by solution that we now turn our attention.

6

Solutional landforms

Almost a century ago, the French geologist, Hubert (1920) described large caverns, not in limestone but in quartz sandstone in western Africa. And karst landforms in sandstone were demonstrated again – albeit accidentally – in 1963, when E. S. Hills published in *Elements of Structural Geology*, a striking air photo – taken during a wartime intelligence flight – of what he described as limestone tower karst in northern Australia. For in that same year, it was shown that this terrain, now known as the Ruined City, was not in limestone, but in siliceous quartz sandstone (the error was corrected in the 2nd edition of Hill's book). Since then, an increasing number of similar, and even more striking examples have been documented worldwide (Fig. 6.1). Yet these features have often been designated only as 'pseudokarst', that mimic the true karst developed by solution of limestone and dolomite, because the features developed in sandstones have been thought not to form mainly by solution. Nonetheless, White *et al.* (1966) described dolines and caves in Venezuela as 'quartzite karst', and later observations in South Africa led Martini (1979) to argue that such landforms truly are karst because solution is the *critical* process in their formation. Observations in northern Australia led Jennings (1983), an expert on limestone geomorphology, also to abandon the term pseudokarst, because of the critical, though not dominant, role of solution in the development of caves and towers in highly siliceous rocks. Mikulas (2008b) has also rejected the term pseudokarst, and proposed the term porokarst to replace it, because of the important role of porosity in the weathering of sandstone. But this term does not adequately convey the dominance of fracture systems in the movement of water through many sandstones. Both pseudokarst and karst are used for very similar features (see for example the range of papers in the journal *Nature Conservation* vol. 63, no. 6, 2007), but many geomorphologists now consider the term pseudokarst no longer acceptable for solutional forms on sandstones and quartzites (Doerr and Wray, 2004).

It is generally accepted that in karst terrain, surface drainage gives way largely to subsurface circulation (Dreybrodt, 1988). Moreover, Mainguet (1972) commented

Fig. 6.1 The labyrinth of corridors and towers cut in the strongly jointed Kombolgie Sandstone, Arnhemland, northern Australia

that movement of water through sandstone not only commonly removes material by solution, but also by mechanical transfer of fines (*crypto–erosion*). She considered this to be proven by the development of limonite-clay lobes where seepage emerges from sandstone rock walls. She further commented that major streams can deepen linearly as they erode headward along fractures, but small streams have little capacity to deepen because they lose so much of their discharge to subsurface flow down joints and fractures. Thus groundwater movement, as well as surface flow, must be important in the development of landforms in stream headwaters on sandstone. Subsequently, R. W. Young (1986) argued that tower features in the Bungle Bungle Range of Western Australia are karst, despite their apparent lack of subsurface drainage, because solution of quartz cement in the sandstone is critical to their development. What then seemed controversial is now widely accepted (Wray, 1997b). Arenization has been identified as the process by which true karst develops in quartzose sandstones (see Chapter 5), and spectacular karst terrains have been described from many countries. Indeed, over the last few decades, the study of solutional weathering and the development of karst features has become the primary research thrust in sandstone geomorphology.

Cavernous weathering

However, solutional weathering is not restricted to karst terrain. Solution is an important process also in landform assemblages shaped by mass failure and

Fig. 6.2 Cavernous weathering at the base of a curved face in the Grampians, Victoria, Australia. Note the control of the face by sloping joints, the tessellation even on steep faces and the collapse of blocks from the cavern roofs

stream erosion. Many sandstone landscapes have stepped slopes, where long walls of cliffs alternate with gently sloping benches. On benches that are not covered completely by soil or vegetation, and in stream beds, a fascinating array of small-scale landforms characterize the rocky outcrops. These include pits and hollows, networks of shallow gutters, rock doughnuts, boxwork patterns of indurated joints, potholes interconnecting different levels of erosion, and oddly shaped masses projecting up from the rock surface. The processes that form these small features vary, but it is our contention that chemical – rather than mechanical – weathering is almost invariably the dominant agent. On cliff faces, the most common features are pits and caverns (Fig. 6.2). Caverning is also widely known as *tafoni*, a term derived from a Sicilian word for window, because growth of the cavern into a block may lead to breaching of the block's surface. Some writers (e.g. McGreevy, 1982) prefer to restrict the term to honeycomb weathering where the hollows within the rock are at centimetre rather than metre scale. Twidale (1982), however, follows the more generally accepted broader usage of tafoni for shallow caverns and hollows, and uses the term 'alveoli' for the very small pits of honeycomb weathering. Honeycombing may be found within caverns and the weathering processes responsible for both forms are clearly related.

Many sandstone faces and boulders are pitted or caverned by hollows eating up into the rock. This caverning is by no means confined to sandstone but is common in granitic rocks (e.g. Twidale, 1982) and is found in a variety of rock types especially in arid or coastal environments (Blackwelder, 1929; Conca and Astor, 1987). It is not however characteristic of highly soluble lithologies such as limestone or dolomite (e.g. Butler and Mount, 1986). There are, of course, instances where honeycombing or pitting of sandstone is due simply to removal of easily soluble cements such as calcite or dolomite. Frye and Swineford (1947) demonstrated the importance of carbonate weathering in the Cretaceous sandstones in central Kansas, and it seems also to be the case in massively pitted sandstone of the Chongan Formation in eastern China (Construction Department of Hunan Province, 2007). But it is by no means limited to sandstones rich in carbonate.

In some cliffs, large expanses of the rock show surface weathering and hollowed-out caverns of metric dimensions; in others, the faces are pitted between closely spaced bedding and vertical joint planes to give a pattern of regular shallow cells (e.g. Pouyllau and Seurin, 1985). In the southern Sydney Basin of New South Wales, Australia, sandstone clifflines are generally caverned. Analysis of panoramic photographs of the Nowra and Snapper Point Sandstones in the Clyde Valley, and the Narrabeen and Hawkesbury Sandstones in the Blue Mountains, show that the areas of cliff faces occupied by caverns vary from 3–28%, with an average of 11% (Young and Young, 1988).

Cavernous weathering is probably as characteristic a feature of sandstones as fluting is of limestones, but it is certainly not uniformly or universally developed. For example, the Torridonian Sandstones of northwestern Scotland show negligible caverning; and the large caves which characterize many of the sea cliffs of the Old Red Sandstone in northeastern Scotland are due to joint widening and block collapse, rather than cavernous weathering. In the southwestern USA where sandstones are prominent in the landscapes, the sheer faces of Coconino and Supai Sandstones cropping out along the sides of the Grand Canyon are sparsely caverned; the tall spires of De Chelley Sandstone at Monument Valley in Utah are unmarked by caverns; the slickrock surfaces of the Navajo Sandstone are pitted by solutional hollows known as 'waterpockets' and the cliffs have many arched hollows due to block collapse, but cavernous weathering is uncommon. It is, however, beautifully developed in the Aztec Sandstone of The Valley of Fire in Nevada (Fig. 6.3). Why this difference?

In many respects, the Navajo and Aztec Sandstones are similar. Both are fine-grained, quartzitic, porous and often cross-bedded. The light-coloured beds of the Aztec develop slickrock surfaces like those on the similarly coloured Navajo. The red, hematite-cemented beds of the Aztec are caverned, although there is no evidence of iron mobilization promoting the caverning (see Conca and Rossman,

Fig. 6.3 Cavernous weathering in the Aztec Sandstone, Nevada, USA. Small pits along the cross-beds indicate the role of water seeping along these partings in the rock

1982). The difference may lie in contrasting patterns of flow through the rock masses. The extensively caverned Aztec Sandstone is low in the landscape; whereas the uncaverned Navajo stands in high cliffs or as extensive dipping slickrock surfaces. Graf *et al.* (1987) commented that pervasive joint systems on the Colorado Plateau inhibit throughflow processes by allowing rapid infiltration to groundwater. The Navajo Sandstone is an important aquifer of high transmissivity, with water being stored both in intergranular spaces and fractures. At its base, close to the contact with the less permeable Kayenta Formation, carbonate cementation and pore space infilling by kaolinite reduces permeability and forces groundwater outflow. This leads to sapping, niche development and block collapse, rather than to cavernous weathering. Thus rapid flow down joints and through very porous strata appears to promote seepage emergence along bedding planes (as in the Navajo Sandstone); whereas slower percolation may promote caverning of cliff faces (as in the Aztec Sandstone). Howard and Kochel (1988, p. 27) made the perceptive comment that a 'surface protected from surface runoff is a necessary condition for tafoni and alcove development'. Where surface wash crosses an outcrop, there is no cavernous weathering.

Moisture flow and silica solution

Certainly, at the small scale, the rate of moisture flow through the rock controls caverning (Conca and Astor, 1987). Iron-stained silica coatings on the rock surface of dolerite and sandstone blocks in Antarctic dry valleys are deposited by flow over the surface. They reduce the permeability of the rock and reduce inflow from the surface. Also, and importantly, they divert capillary flow within the rock away from the coated surfaces and towards uncoated surfaces where evaporation may occur. Increased moisture flux in a solutional weathering process will increase weathering by increasing the chemical potential of water, especially at grain boundaries. The shape of caverns within weathered boulders follows the shape of lines of matric potential (negative porewater pressure) due to capillary flow and water film effects. Spatial differences in the permeability will cause spatial differences in the moisture flow through the rock to produce corresponding morphologies. Antarctica and arid regions in general have a greater variety and complexity of weathering forms because the relative absence of fluvial activity and the low total water budgets produce more complex moisture fluxes that act over longer time spans than in more humid regions (Conca and Astor, 1987).

It is the movement of water through the rock that initiates and expands caverns. It has long been recognized that active weathering within caverns involves disintegration of the rock, so that loose grains fall from the surface and accumulate on the floor (Fig. 6.4). There is great variety of form, but consistently, the outside of the caverned area is more coherent than the inner surface. Flaking can also

Fig. 6.4 Caverns on the sea-cliffs cut in Hawkesbury Sandstone near Sydney, Australia, show dribbles of sand (see inset) on actively eroding faces

occur, adding thin platey rock fragments to the floor sediments. Debris may be (but often is not) removed by wind, and – contrary to much popular opinion – sand-blasting is not the means by which caverned niches are formed (Blackwelder, 1929). Rain and/or animal movements may remove floor sediments, but in many protected sites, sediments accumulate to considerable depths. For example, in some caverns in southeastern New South Wales, the floor sediments are over a metre deep, deep enough to yield evidence of several thousand years of Aboriginal occupation (Sullivan and Hughes, 1983). Many caverns have friable plates of rock projecting from the interior walls, especially where weathering has concentrated along cross-bedding planes. Many also have small tubes puncturing the rock on the back walls, patches of honeycomb weathering on the interior walls and a relatively hard lip or overhang at their entrances.

Caverns usually have sections of their interior walls where the processes of cavern extension are inactive or dormant. On these sections, there is no loose granular material ready to fall and the rock surface becomes colonized by lichen or algae, or perhaps stabilized by inorganic precipitates. Where cavern extension has breached the cavern wall, creating a window near the top of the hollowing block, the interior wall near the window is almost invariably inactive. Similarly, where cracks or joints allow rapid drainage of water through the cave, caverning appears to become inactive. Development controlled by flow vectors within the rock and evaporation from surfaces that are not case-hardened or coated, as proposed by Conca and Astor (1987), accounts for these characteristics adequately.

Expansion of caverns up into the rock mass led some workers (e.g. Dragovich, 1969; Turkington and Paradise, 2005) to suggest that the microclimate within the cave is an important factor. In such cases, the expansion of hydrated clays might be the primary mechanism causing rock break-up, especially in clay-rich sandstone. Such a process is analogous to the expansion thought to be caused by the growth of salt crystals within the rock, from salt-laden solutions moving through the blocks, and postulated particularly for salt-rich environments such as arid and coastal areas.

There is no doubt that caverning and honeycombing are better developed and more common in salt-rich environments than in humid environments, and a voluminous literature attests to the efficiency of salt solutions in causing rock disintegration (e.g. Goudie, 1986; Smith *et al.*, 1987). Using evidence from experiments on mica schist, Davison (1986) proposed a supplementary role for temperature. Simulation of salt weathering produced most debris in cold conditions (simulating Antarctic temperatures), less debris in hot conditions (simulating Tunisia) and – perhaps surprisingly – least in temperate conditions (simulating southwest England). Certainly the most cited cause of salt weathering is the pressure exerted by crystal growth. However, we believe that the action of salt is related more to its influence on rates of solution than to any mechanical

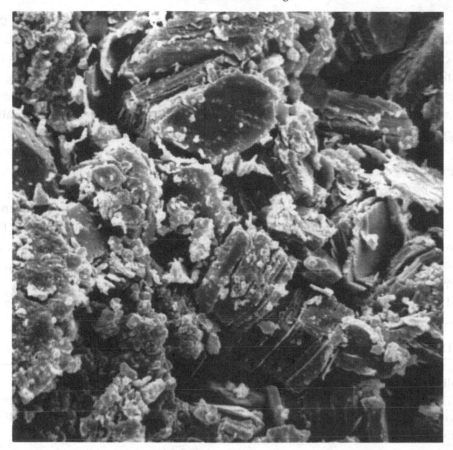

Fig. 6.5 Clay crystals (SEM image, 2000×magnification) from the matrix of Hawkesbury Sandstone, within an active cavern, showing etching and disintegration

pressure it exerts, and that caverning and honeycombing are essentially solutional and not mechanical processes.

A.R.M. Young (1987) suggested that the role of salt both in arid/coastal environments and in humid environments is to increase the rate at which silica dissolves. A high concentration of salt decreases the equilibrium solubility of quartz, but the equilibrium solubility is less important in geomorphic processes than the speed with which silica moves into solution, into water percolating down through the rock. The silica dissolved may come from the grains, but it is the loss of cement that seems to be the main cause of cavernous weathering. Intense etching of the clays between quartz grains can be seen in SEM images of samples from active caverns in sandstones of the Sydney Basin (Fig. 6.5). This suggests that silica removal from the clays may occur more rapidly than removal from the quartz, whose grains are larger and have lower surface area:volume ratios and fewer boundary contacts than the stacked clays. The floor sediments from the caves in Hawkesbury Sandstone include etched quartz grains, and also aggregates of clay etched on the margins but sometimes still displaying stacked structure, testifying to both silica solution and destruction of cementation and bonding in the rock. Mainguet (1972), in her discussion of sandstone matrix and cement, commented that dissolution of quartz is greatest in argillaceous zones. She also suggested that percolating water may wash clay out of voids in the rock, a physical process akin to lessivage in soils. Certainly, some iron-oxide-rich tufas contain clays; and the skins on towers in the Bungle Bungles apparently form partly by clays shifting out from the interior of the towers to coat the surfaces. The variations in iron oxide concentrations within many caverns in sandstones raise the possibility of iron mobilization also being important, as proposed for the Hawkesbury Sandstone (A. R. M. Young, 1987) and in plutonic rock (Conca and Rossman, 1985). The evidence suggests however that silica loss and clay weathering are more important in cavern development.

Two aspects of weathering of the rock need to be considered (Conca and Rossman, 1985):

- The granular framework of the rock, with close-packed geometric arrangement of grains. This can be disrupted as micro-fractures develop, and then as material is dissolved and removed along these fractures.
- The chemical component such as bonds within mineral lattices or between grain surfaces. Alteration of primary minerals, breaking of bonds between grains, crystallization of secondary minerals and other chemical changes affect this component.

The two aspects are inter-related. For example, weathering of feldspars to clays, and ingress of water along micro-fractures may cause expansion pressures within the granular framework, and development of more fractures. This explains the prominent control of weathering in sandstone by variations in the structure and orientation of bedding lamellae (Young and Young, 1992).

Case hardening or core softening?

The presence of a relatively hard layer on the exterior walls and around cavern entrances has been attributed by many workers to case hardening. This is the precipitation of secondary minerals in the pore spaces of the original rock, a process which increases the hardness and reduces the permeability of the affected zone. For example, case hardening by iron oxide minerals is common in the sandstones of the Sydney Basin, New South Wales; and Conca and Rossman (1982) described case hardening in the Aztec Sandstone, Valley of Fire, Nevada, by infilling of pore spaces with kaolinite and calcite. Scanning electron

Table 6.1 *Schmidt hammer readings from a cave in the Nowra Sandstone, southern Sydney Basin, Australia*

Section of the cave	Adjusted rebound reading
Case-hardened lip at the entrance	30–54 (mean 45)
Sandstone behind the hardened lip	28–37 (mean 33)
Interior of fretwork hollows	12–28 (mean 20)
Rims of the fretwork	18–33 (mean 21)
Back wall of the cave	14–40 (mean 28)

micrographs show the interiors of caverns at the Valley of Fire to have rounded etched grains and high porosity, in contrast to the case-hardened rims where porosity is much reduced by surface coatings of kaolinite and calcite on the sand grains. However, when discussing caverning of plutonic rocks in Baja, California, Conca and Rossman (1985) suggested that core softening – not case hardening – is the major process. The interior of the boulders showed evidence of considerably more chemical weathering and kaolinization than the exterior, or than a nearby, unweathered exposure. In this instance, the weathering was due mainly to selective solution of iron-rich layers in biotite, and to chemical weathering concentrated along grain boundaries and cleavage planes in feldspars. Some of the iron then migrated to the exterior walls; and there were more frequent hematite aggregates and higher Fe^{3+}/Fe^{2+} ratios in the rock of the exterior walls.

These weathering changes meant that the rock in the interior was more friable than that on the exterior. For the quartzose Nowra Sandstone, near Nowra, New South Wales, the relative hardnesses of rock within one cave are shown by the Schmidt hammer values in Table 6.1, with the readings adjusted for the variable angle of the hammer.

The differences in hardness indicate that this cave was core softened. The fretwork was softer than the back wall or the rock behind the case-hardened lip. Conca and Rossman (1985) found similar differences to be at least on the scale of an order of magnitude. Scanning electron microscope studies of caves in Hawkesbury Sandstone near Wollongong, New South Wales, illustrated the reasons for such difference. Near the lip, the quartz overgrowths were sharp-sided and closely interlocking; the clay cement showed stacked hexagonal structure and almost filled the voids between quartz grains. On the active part within the cavern, etching of the overgrowths had partially rounded the quartz grains; the clays no longer showed the typical stacked structure but were etched and had tubular hollows through them. Thus, the porosity of the actively weathering zones was higher; and the flux of moisture, and thence the rate of weathering of the interior, would exceed the rates for the exterior walls.

Fig. 6.6 Huge caverns etched up into the cliffs of Nowra Sandstone on Mt Horton, southern Sydney Basin. Note the cross-bed plates projecting within the cavern

Caverns hollow up into the blocks in which they are initiated, not because of the cavern microclimate, but because of the inter-related high permeability and high rate of weathering in the boulder interior. It is seepage through the rock – not expansion forces in the pores of the rock near the rock surface – which is responsible for cavernous weathering. The role of permeability is also indicated by the fact that caverns are preferentially developed in the conglomeratic or pebbly beds in upwards-fining sandstone sequences, or above bedding planes in sandstone of more uniform texture. Within caverns, it is indicated by preferential erosion along cross-bedding planes (Fig. 6.6). In some cases, erosion of low deep caves may cause collapse of the caverned block. As noted earlier, stalactites spread across cavern roofs show that silica moves in seepage water through the blocks of sandstone, and not just down fractures. Extension of the cavern occurs by granular disintegration of the active surface after solutional weathering has destroyed interlocking and cementation within the micro-framework of the sandstone.

The role of salt

What then is the role of salt? Experiments using salt solutions demonstrate breakdown of sandstones and photographs show abundant salt crystal growth on the flaking surfaces (e.g. Goudie, 1986; Smith *et al.*, 1987). However, neither the photographs under optical microscopy in Smith *et al.* (1987), nor those under scanning electron microscopy in that and other papers (e.g. Mustoe, 1983;

McGreevy and Smith, 1984) depict the salt crystals supposedly exerting the pressures to mechanically force grains apart. However, they do show etching of quartz grains (especially at grain boundaries or between overgrowths and the detrital grain underneath), pitting of clays so that they appear almost tubular (McGreevy and Smith, 1984) and disaggregation of grains along micro-fractures (Smith *et al.*, 1987). Butler and Mount (1986) commented that corrosion pits in varying lithologies show etching along grain boundaries of quartz and feldspars, due to dissolution, and noted that salts, where present, do not fill pore spaces. Disaggregation is not then mechanical. Figure 5.3 shows a salt crystal resting on the highly etched grains, not filling voids between grains. All these features are compatible with a solutional chemical role for the salt solutions used in the experiments, rather than a mechanical crystallization or hydration pressure.

If we then look at honeycomb weathering in a coastal area, the influence of solutional etching can be seen at all scales. For example, study of the quartzose Nowra Sandstone near Jervis Bay in New South Wales showed that in the specimen at hand, the rock is pitted to a depth of about 1 cm by semi-circular hollows with rounded rims. The quartz grains are frosted and lack the glassy lustre of grains from unweathered rock (Fig. 6.7). Under SEM, the honeycombed

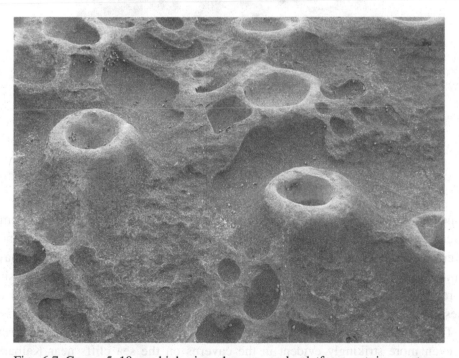

Fig. 6.7 Cones 5–10 cm high rise above a rock platform cut in quartzose Snapper Point sandstone in the southern Sydney Basin. The platform has been lowered by solution, leaving the cones rising from it. The reason for apparent hardening of the cones is probably silica induration

Fig. 6.8 Tufa column stretching down the face of a cavern in the Tumblagooda Sandstone, Kalbarri, Western Australia. The calcareous tufa has formed from seepage out of the base of the pale sandstone. The dark beds are claystones and clayey sandstones

appearance persists even to magnifications of 1300×, with tubular and semi-circular etch pits throughout the specimen. Few sharp edges survive on the quartz overgrowths; and the platey clay structure is etched so greatly that the clays have an almost flower-like appearance. No salt crystals were observed, but dispersive X-ray spectral analysis showed Fe, Cl and K associated with the Si and Al on these etched clays.

The solutional transport of material in association with cavernous weathering is even more strikingly evident in the caverns on the sea cliffs near Kalbarri, Western Australia. The cliffs are cut in the Tumblagooda Sandstone, a quartzose and hematite-rich sandstone, which is characterized by caverning particularly in those facies which are of marine rather than fluvial origin. In this semi-arid and coastal environment, salt is presently being blown onto the cliffs and thick wedges of salt form at the outlets of seepage along bedding planes. Whether the salt plays a mechanical role is doubtful. SEM images indicate strongly that the role is solutional (A. R. M. Young, 1987). And it is not only salt which is mobile on the sea cliffs. Seepage over the lips of overhangs leads to precipitation of tufas that may develop sufficiently to form a column from the base of the overhang to the lip (Fig. 6.8). The source of some of the material in these tufas may be the Pleistocene aeolinites that lie on the cliffs above the Tumblagooda Sandstone, while the deep red–brown colour of the tufas indicates that substantial iron movement is also occurring. Thus, in this sandstone environment as well as in the humid temperate environment of southeastern Australia, the seepage of water through the rock mass and down joints and partings – and the resultant movement of materials in solution, and their subsequent re-precipitation – is crucial to the development of weathering features.

In summarizing the extensive scientific literature on the development of cavernous weathering, Turkington and Paradise (2005, see especially Fig. 5) have emphasized the interplay of salt crystallization, dissolution of the cementing agent, and the relocation of clay minerals. This interplay leads to salt efflorescence, granular disintegration, flaking and scaling, loss of cohesion and increased porosity that in turn result in the deepening of the cavern. They have also argued for a concurrent stabilizing of the surface of surrounding rock by the deposition of minerals relocated from the sandstone, together with the addition of allochthonous materials, and the growth of biological materials. Our concern is that insufficient emphasis has been given, in this otherwise very useful summary, to the critical role of solution in the breakdown of the cement and the clasts, and also to the chemical rather than mechanical role of salt.

Small surface features

Solution basins

The difference between shallow caverns that extend back and up into sandstone, and basins that extend down into it, is more than a simple matter of orientation, for the pitted and loose surface layer of caverns contrasts with the much smoother and more cohesive floor of basins. For example, on sea cliffs in both the Nowra and Hawkesbury Sandstones of the Sydney Basin, pitting is confined to sloping surfaces, projecting plates and small ridges in the rock. Wherever water can lie for some time after rain or wave washover, the rock surface is generally almost flat with smooth-sided and very shallow hollows; there is rarely a covering of plant debris, or more than a few millimetres of loose sand. There is a sharp break between these shallow, smooth-floored and generally circular ephemeral basins and the very pitted ridges around them. The basins often drain down to another level, where the same pattern is repeated. In some cases, the breaks of slope between basins are not abrupt, and sinuous runnels may connect the pools. This pattern of small basins connected by runnels is also the norm at inland sites in the Sydney Basin. However, away from the coast, the small ridges between the basins are rarely pitted and rough, but rather they are smoother and lichen-covered. Moreover, the basin floors are either bowl-shaped or flat, and are often covered by water and plant debris.

Basins can form on many rock types (Hedges, 1969), but are common surface features on sandstones virtually worldwide (Wray, 1997a), as the following brief list of examples shows:

- In North America, on the Colorado Plateau (Howard and Kochel, 1988), and on crests of ridges on gently dipping quartzite in the foothills of the Blue Ridge in the Appalachians (Reed *et al.*, 1963).
- In South Africa, on the Clarens Formation (Cooks and Pretorius, 1987).
- In Australia, both in temperate environments, such as the Sydney Basin, the drier Grampians of western Victoria, and tropical environments of northern Australia (Wray, 1997a).
- In Britain, albeit with varying degrees of development, in several sandstone terrains (Robinson and Williams, 1994).
- In continental Europe, in the Fontainebleau region of France (Fränzle, 1971; Thiry, 2005); and the Bohemian Basin (Varilova, 2002; Mertík and Adamovic, 2005), and Polish Carpathians (Alexandrowicz, 1989).
- In South America, in the Roraima region (Pouyllau and Seurin, 1985).

The plan forms of individual basins are commonly near-circular or oval, but sometimes irregular. Where several basins have formed in close proximity, they may coalesce by erosional breaching of their intervening walls to form larger, irregular or amoeboid basins. Generally there is no obvious control by joints or cross-beds. The long axis of basins, best seen in the elliptical examples, may be guided by joints, but typically has no preferred orientation. A weathering residuum of sand grains or small pebbles, and often a thin layer of silt, moss, lichen or plant litter accumulates in the floors. This may be washed or deflated out of the basins, but it may accumulate in the largest basins to a significant depth to form a sandy soil and even support shrubby vegetation (Fig. 6.9). Retention of moisture in such basins may enhance the enlargement process. Some basins have well-defined outlet notches or low points on their rims, especially those on gently sloping surfaces; but a large proportion of those on near-horizontal surfaces have no definite outlet. Where spillways or runnels are found, usually on sloping rock

Fig. 6.9 A basin in a tessellated outcrop in Monolith Valley, southern Sydney Basin

pavements, they are frequently well defined and indicate the path of water overflowing the basins. Several basins may nest within a larger one.

Little dimensional analysis of sandstone rock basins has previously been conducted. Surprisingly, given the numerous references to basins in the karst literature, it seems that very few detailed data sets have been published for basins in limestone either. Available reports generally indicate that sandstone basins range from several centimetres to several metres in both length and width, and may be up to several tens of centimetres deep. Very large basins are sometimes found, with dimensions measured in metres, with some extreme and unusual examples reaching sizes of several tens of metres (Netoff and Shroba, 1993, 2001). Dimensional analyses of sandstone basins are available from Fontainebleau in France (Fränzle, 1971), the Colorado Plateau (Schipull, 1978), and from the Clarens Formation of northwestern Transvaal (Cooks and Pretorius, 1987). Unsurprisingly given the basins' near-circular shapes, there are strong positive correlations between basin width and length, with r^2 values of 0.99 (Fränzle, 1971), 0.95 (Schipull, 1978) and 0.92 (Cooks and Pretorius, 1987). Individually, the relationships between length and depth are also, in at least two cases, very strong (0.44, 0.81 and 0.97, respectively). In the Sydney Basin, harder sandstones, being more resistant to general surface denudation, display more and better-developed basins than the region's softer sandstones. As elsewhere, they are generally much wider than they are deep. Statistical analysis showed comparable sizes to the basins studied by Fränzle (1971), Schipull (1978) and Cooks and Pretorius (1987), with similar relationships between the size parameters, and also with highly significantly positive correlations in basin width and length (0.84).

Limestone basins are generally attributed to solutional processes, and similar processes may be important on carbonate-cemented sandstones (Robinson and Williams, 1994). Basins in arkosic sandstones, like those of the Colorado Plateau, may be largely the result of removal in solution of calcite cement (Howard and Kochel, 1988). However, the formation of basins in highly siliceous sandstone is considered more problematic (Robinson and Williams, 1994), although most authors agree that the dominant processes are chemical or biochemical dissolution facilitated by standing rain water in the basins. Basins in the Fontainebleau sandstone show two types of silica behaviour – dissolution at the base and secondary silica deposition on the indurated rims (Thiry, 2005). Thiry believed that, because of the steep and unstable slopes on which some of these basins occur, the rate of weathering must be relatively fast.

Temporary retention of water seems to be an important control. The development of basins on sandstones in Britain varies largely with the porosity of sandstones, with basins more common in less pervious rock (Robinson and Williams, 1994). Basins in sandstones near Sydney are often most common on upper to mid-slope positions, with many located on the crest of sandstone pavements, and some even on the crest of isolated boulders. However, while basins are usually found on flat to gently inclined surfaces, they also occur in chains connected by shallow runnels or spillways cascading down steep slopes. In Utah, they occur on sandstone slopes as steep as 30° (D. Netoff, *personal communication*, 1996); and in central Australia, they occur not only on the crest of Uluru but also on the flanks sloping at more than 40° (see Fig. 4.8).

The salt crusts that are found in most basins along sandstone shorelines have generally been taken as proof of the dominant role of the mechanical disruption of the rock by the pressure of crystallization. However, evaporation of sea water does not lead to the growth of salt crystals directly on or in the sandstone, but rather as a layer separated from the rock by the remaining water. Even when the remnant of the water is evaporated, the crystal growth is primarily upwards towards the minimal confining pressure of the air above, not downwards into the rock. Sand grains trapped within the salt crust are more likely to be the product of prior weathering or have been washed into the basins during high tide. We suggest again that the role of salt is primarily chemical rather than mechanical.

Runnels

Small channels formed by weathering as water flows across carbonate rocks have been studied in much detail, but far less attention has been given to their occurrence on sandstone. Bögli's (1960) classification for limestone runnels distinguished between rillenkarren – which are runnels formed by the direct action of sheetflow of rainwater – and a larger group – formed by channelized water flowing across a surface from an external source – which he termed rinnenkarren, rundkarren, and decantation forms.

Rillenkarren are small, V- and U-shaped suites of straight channels of regular form and dimension that head at the crest of steep bare rock slopes and extinguish down slope. They are common on some limestones, but virtually unknown on sandstones. Robinson and Williams (1992, p. 426) described 'a strange micro-topography of sharp-crested ribs or ridges' on quartz sandstones of the Atlas Mountains in Morocco as bearing 'a somewhat similar appearance' to rillenkarren. They were equivocal, however, as to whether these features were rillenkarren *sensu stricto*. Rather than being the product of sheetwash, they may in fact have developed beneath winter snow drifts by carbon-dioxide-charged melt waters, or have formed by percolating water charged with organic acids beneath a former soil cover. The lack of rillenkarren on most sandstones is probably a result of the coarse granular structure of the rock, as rillenkarren often do not form on coarse-grained limestones, marbles or dolomites (Ford and Lundberg, 1987).

Rinnenkarren are much bigger than rillenkarren, generally 12–50 cm wide, and begin where sheetflow down a slope breaks into linear streams (Fig. 6.10). They are separated by distinct interfluves, have sharp channel rims and rounded bases, and increase in depth and width downslope with increasing catchment area. Rundkarren in limestone generally form beneath a cover of soil or other material (Bögli, 1960; Ford and Lundberg, 1987) and are much more rounded in section than rinnenkarren. They both commonly display dendritic or meandering plan forms on low angle surfaces, and on higher slope angles tend to be sub-parallel. They may deepen downslope, and their lengths are variable and dependent on the volume of water available, length and gradient of slope, rock texture and amount

Fig. 6.10 Rundkarren and decantation runnels formed by solution as seepage from swampy vegetation flows down a sloping rock face in the Shoalhaven River catchment, Sydney Basin

of cover removed. There are also runnels formed by decantation or overspill processes where the water is supplied either perennially or intermittently from an upslope store, such as a patch of soil or vegetation, rather than from direct precipitation (Ford and Lundberg, 1987).

Robinson and Williams (1994) reported fluting on many steeply sloping sandstone surfaces in Britain and continental Europe. These long, parallel vertical flutes are reminiscent of wandkarren in limestone. They noted that they are quite common on impervious sandstones (such as the Millstone Grit and Fell Sandstone), but uncommon on very porous sandstones (such as the Lower Cretaceous sandstone cliffs in Kent and Sussex or on the Devonian Old Red and Torridonian Sandstones). They also noted that many Neolithic standing stones are grooved by vertical flutes, suggesting a relatively rapid formation, certainly within the last 3200–4000 years. Similar conclusions were drawn by Self and Mullan (2005) for well-developed decantation runnels or flutes on the Duddo Stones, a set of standing stones of a Neolithic stone circle.

Many sandstones and quartzites display excellent examples of various runnels, but do so more rarely than granites. Robinson and Williams (1992) described shallow channels or gutters up to 100 mm wide and 10 mm deep, inset into sloping sandstone pavements in Morocco. In South Africa, Marker (1976) reported sequences of runnels draining basins on the quartzites and metamorphosed sandstones of the Transvaal. White *et al.* (1966), Pouyllau and Seurin

(1985) and Piccini (1995) reported frequent grooves and intermediate ridges with a groove-to-ridge relief of 25–50 cm, on the quartzites and sandstones of the Roraima. De Melo *et al.* (2004) noted the widespread occurrence of 'lapiés' or runnel forms on the sandstones of the Vila Velha State Park, Southern Brazil. And Howard and Kochel (1988) documented extensive systems of runnels on the Navajo Sandstone of the Colorado Plateau.

Runnels are also very common on a number of Australian sandstones. They occur in the tropical environments of Arnhemland, the Kimberley region and northeastern Queensland, through the semi-arid sub-tropics of the Carnarvon Range in central Queensland, to the southern temperate environments of the Sydney region and the Grampians. In contrast to those on most limestones, runnels in sandstone of the Sydney region (Wray, 1996) often display characteristics of both rinnenkarren and rundkarren along their length. Some sections of a runnel may be well defined with sharp rims and V- or U-shaped cross-sections (like rinnenkarren); while other parts of the same runnel may be much more rounded and less distinct (like rundkarren). These changes may even occur several times along a single runnel. Many runnels are also flat-floored, not rounded as in the classic limestone forms; and the base commonly changes from flat to rounded or irregular many times along an individual runnel. Runnels in the Grampians have similar forms. Decantation runnels are probably just as common as those fed by direct precipitation runoff. On very steep surfaces, sequences of parallel sharp-edged rills that shallow downslope (like wandkarren) are sometimes seen. These are excellently developed on the steep sides of the towers at Monolith Valley south of Sydney (see Fig. 4.6), and also at Wonderland and Mt Stapylton in the Grampians.

Although grains detached from sandstone by weathering are transported by surface flow down runnels, the runnels themselves appear to have been formed dominantly by solution. This seems to have been very much the case in the Sydney region, where the edges of runnels are commonly coated by layers of silica precipitated from water running down the sandstone surface. These coatings are particularly well developed where runnels carry water draining from patches of swampy vegetation. Iron-rich skins that coat some runnels also indicate a solutional origin for these features.

Caves

Whereas cavernous weathering is essentially a surface feature, many caves or caverns penetrate much more deeply into sandstone. Caves and other underground drainage systems are now probably the most widely described feature of quartzose karst; and while none of the known caves in sandstone or quartzite attain anywhere near the size, length or depth of the larger limestone cave systems, they are nonetheless comparable in size to the vast majority of smaller limestone caves.

Dreybrodt (1988) noted that flow through micro-fissures in limestone (with diameters less than 0.1 cm) is very slow and is laminar. Substantial percentages of the total solution in a limestone, and of the water stored within it, are associated with these very small fissures. Once they are enlarged to pipes with diameters of about 0.1 cm, the flow becomes turbulent and the rate of flow increases dramatically (perhaps 1000-fold). Both these changes trigger a jump in the diffusion coefficient and a resultant jump in the rate of solution of the pipe walls. In turn, this draws flow into the pipes from adjacent fissures in which flow is still laminar, and cave systems develop along the more efficient drainage pathways. Jennings (1985, p. 67), when discussing limestone terrain, noted that water 'moves anisotropically through narrow fissures and large caves which ... could be regarded as separate aquifers.' Thus, there may be no single water table in karst terrain, but rather independent conduit systems and rest levels. This suggestion echoes the remarks of Mainguet (1972), who referred to multiple horizontal levels of flow (*nappes*) within sandstones and commented that subterranean flow is not so much a sheet but a network. The less permeable layers, by limiting free infiltration of rainwater, generate *nappe* aquifers. These impeding layers have a wide range of degrees of lowered permeability and allow greater or lesser infiltration. The aquifers feed one another, from higher to lower, via the generally sub-horizontal dips of the strata, passage to other sandstone strata, or – the most frequent case – use of the outlets comprised of *diaclases* (large-scale fractures). Only the *nappes* near the surface of the soil are fed directly by precipitation. Thus 'the circulation of water, whose speed is already reduced by the very nature of the material, is further limited by the veritable labyrinth that the water must flow through' (Mainguet, 1972, p. 101, transl. A. R. M. Young).

Caves in the Roraima

The most commanding cavernous quartzite and sandstone region in the world has developed within the isolated mesas (*tepuis*) of tropical Roraima in Venezuela, where the frequency and sizes of caves far exceed that reported from any other region in the world. As this is an area of high rainfall, much of the surface runoff cascades over the massive quartzite cliffs. Water that proceeds underground via a myriad of fractures and sinkholes finds its way by large and intricate, often joint- and bedding-controlled cavern systems (Fig. 6.11) formed by the mechanical removal of sand after partial solution of the siliceous cement (Briceño and Schubert, 1990; Piccini, 1995; Doerr, 1999), to re-emerge on the vertical walls of the *tepuis*, up to several hundred metres below the summits.

Fig. 6.11 An active cave in the Roraima quartzite, Kukenan *tepui*, Venezuela.
(Photo: S. Doerr)

Major discoveries of these caves were made during the early 1970s. A relict
complex passage system over 400 m long with phreatic tubes up to 20 m in
diameter was reported within Cerro Autana, 650 m up this 800-m high mountain
(Colveé, 1973; Urbani and Szczerban, 1974; Urbani, 1976). It was argued that
this cave is significant not only for its large size, but because it is also the oldest
known cave in the world, being thought to have formed during a period of
phreatic activity in the Precambrian (Colveé, 1973); it is thus far older than any
known limestone cave. Urbani and Szczerban (1974) briefly described several
active river caves, one in Territorio Amazonas passing through Guanay Mountain
to emerge 800 m from where it sank, and another resurging over 1000 m from its
sink. The Sarisariñama Plateau in nearby Bolivar State has numerous large
dolines and caves (Zawidzki *et al.*, 1976) in hydrothermally weathered Roraima

Table 6.2 *The longest and deepest sandstone caves and shafts of the world (Martini, 2000b; Carreño and Blanco, 2004; Galán et al., 2004; Audy and Šmída, n.d.)*

Cave system	Length	Depth	Location
Sistema Roraima Sur	10.8 km	72 m	Venezuela, Roraima
Cueva Ojos de Cristal	5.3 km	72 m	Venezuela, Roraima
Cueva Charles Brewer	4.8 km	110 m	Venezuela, Chimantá
Gruta do Centenario	3.8 km	481 m	Brazil
Gruta da Boccaina	3.2 km	404 m	Brazil
Sima Auyán-*tepui* Noroeste	2.9 km	370 m	Venezuela
Cueva del Diabolo	2.3 km	–	Venezuela, Chimantá
Sima Aonda Superior	2.13 km	320 m	Venezuela
Magnet Cave	2.49 km	–	South Africa
Sima Aonda	1.88 km	383 m	Venezuela
Sima Acopán 1	1.38 km	90 m	Venezuela
Sima de la Lluvia	1.35 km	202 m	Venezuela
Sima Menor	1.16 km	248 m	Venezuela
Sima Aonda 2	1.05 km	325 m	Venezuela
Mawenge Mwena	–	305 m	Chimanimani, Zimbabwe

quartzite. Szczerban *et al.* (1977) also reported numerous caves in Bolivar State, the caves and dolines of the Meseta de Guaiquinima being linked by an underground stream nearly 2 km long. Chalcraft and Pye (1984), Pouyllau and Seurin (1985), George (1989) and Briceño and Schubert (1990) also described large active underground drainage systems and abandoned caves in this area.

Table 6.2 lists the largest of the quartzite caves. The longest is Sistema Roraima Sur (recently joined to the adjacent Cueva Ojos de Cristal), which is now known to be more than 10.8 km in length, even though it is only 72 m in depth (Galán *et al.*, 2004). Volumetrically, the largest sandstone cave so far found is Cueva Charles Brewer. This 4.8-km long cave drains an extensive area of the mountain surface, and runs parallel to the top of the plateau 150–200 m below the surface. Many of the passages contain active streams, which even during the dry season carry an estimated 200–300 L/s. The size and included volume of the underground galleries and passages of this cave exceed all other known quartzite caves; galleries and passages are typically 40 m wide, but some reach 60 m. The biggest enclosed space is the Gran Galería Karen y Fanny, with a domed chamber 70 m wide, up to 40 m high and more than 355 m long (Šmída *et al.*, 2005).

The development of karstic landforms of such scale in such poorly soluble rocks is possible only under very particular environmental conditions. In this case, the geologic stability of this area has contributed to this remarkable land-scape development, with weathering having developed the caves in the 100 million years since the Cretaceous (Briceño and Schubert, 1990; Mecchin and Piccini, 1999).

Piccini (1995) stressed the importance of lithologic and structural control in the caves and shafts of Auyán-*tepui,* noting that the caves follow fractures that act as points of concentrated infiltration, and that all show a pattern strongly controlled by the joint sets. While the caves initially form mainly by solution processes along fractures and joints, their enlargement is by erosion and subsequent collapse. Piccini also noted that the morphology of underground passages is simpler than in carbonate rock cave systems, because of control by the bedding planes and the joints. Phreatic passages with sub-circular cross-sections are also commonly found in many of the inactive galleries, and have rounded erosion-surface ceilings with ceiling pockets, probably forming during high meteoric flows when lower passages were flooded with water.

Doerr (1997, 1999) described caves, dolines, sinkholes and karren features from Kukenan *tepui,* and showed that it is the dissolution of both original grains and quartz overgrowths that is responsible for weakening the quartzite, the weakening being followed by piping of loosened sand grains by flowing waters. He concluded that the main control on the formation of karst features in quartzite is time, and that the rarity of karst features is not 'simply due to intrinsic physical parameters (i.e. low solubility), but rather to the rarity of areas that undergo stable weathering under a suitable climate, without disturbance, for a sufficiently long period of time' (Doerr, 1999, p. 14).

Similar karst hydrology and large quartzite caves occur on the nearby Guyana Plateau (Urbani, 1977), and in various regions of Brazil (Corrêa Neto, 2000; Wiegand *et al.,* 2004). In Minas Gerais, there are at least 35 sandstone caves longer than 500 m, some with active subsurface streams. This is possibly the highest known concentration of long quartzite caves known anywhere in the world. All of these caves are attributed to silica solution followed by vadose mechanical removal of loose sand grains to create linear pipes (Corrêa Neto, 2000).

Caves in Africa

The Roraima and nearby regions of tropical and sub-tropical Brazil are indeed exceptional for the sheer number and size of caves and associated karst features. But smaller numbers of active and relict caves in quartz sandstones have been described from most continents. Caves had been reported from thick bedded Upper Cretaceous sandstones of humid tropical southern Nigeria by early in the twentieth century (Wilson and Bain, 1928, quoted in Szevtes, 1989); but it was not until investigations by Egboka and Orajaka (1987), Mbanugoh and Egboka (1988) and Szevtes (1989) that a number of these caves, some with running streams, were accurately described. More than 16 horizontal caves are known, many of which contain active streams. Several are associated with deeply incised canyons, and in most instances bedding planes, joints and other structural features have exercised important controls on water flow and consequent cave development. The longest is Ogbunike Cave, a 350-m long, complex and multi-level sandstone maze that records a long period of cave and canyon development. As in the Roraima quartzites, the cementing material in these Nigerian sandstones is mainly siliceous and ferruginous (Szevtes, 1989).

The abundance of caves and tubes in the sandstone terrain of the arid Tibesti region of Tchad prompted Mainguet (1972) to state that they are as common in sandstone as in limestone. But they are essentially relicts of a once more humid climate. The same seems true of the abundant sandstone caves in the nearby Saharan region of eastern and northeastern Niger (Busche and Erbe, 1987; Sponholz, 1989, 2003; Busche and Sponholz, 1992). A well-developed sub-terranean network of karst passages in the sandstones is proven by water gushing from many tubes in cliff walls after heavy rain. Hundreds of closed scarp-foot drainage depressions are believed to have been active until the Pliocene; and many relict small phreatic caves and tubes a few centimetres in diameter apparently formed during wetter periods in the mid-Tertiary. Although the total number of accessible caves is small, the sandstones of this area are completely riddled with small inaccessible passages, and mineral dissolution, especially of quartz, has been the critical factor in their formation (Sponholz, 2003). These systems began to form millions of years ago, but the lower parts of the system are probably still filled with water. Willems *et al.* (1996) report similar features in the ferriferous Continental Terminal sandstone of west Niger. Large caves are also found in the currently hyper-arid regions of Egypt and Libya (Halliday, 2003).

However, caves in sandstones are by no means limited to regions in the hot wet tropics, or to regions that have had such climates in the relatively recent geologic past. In southern Africa, Marker and Swart (1995) suggested the karst developed beneath the Late Cretaceous–Mid Miocene African Surface before it was faulted and uplifted in the Miocene and Pliocene. The Berlin Cave in the deeply wea-thered Black Reef Quartzite, of eastern Transvaal, consists of two systems that are large by world standards (Martini, 1979, 1982, 2000b). The southern system consists of two large complex dolines, beneath which are at least 17 caves. Bedding planes are the dominant structural control on cave development; nearly all of the caves contain active streams; and water flows in about 70% of the passages even in the dry season. The caves in the northern system are smaller, but have the same general morphology. Another major cave system in the Transvaal, the Magnet Cave, is in the Daspoort Quartzite. With a length of 2.49 km, this is southern Africa's longest quartzite cave, and one of the longest in the world (Martini, 2000b). While Transvaal has the longest quartzite cave in Africa, Zimbabwe has the deepest. At Turret Towers, in the cool temperate Chimanimani Highlands, deep weathering of the faulted Umkondo Group quartzite has resulted in the development of many large jungle-filled dolines, several of which lead into deep caves. The Mawenge Mwena is 305 m deep.

Caves in Europe, North America and Australia

Karst in siliceous sediments seems to be widespread in Russia. Caves, cones, depressions and holes are described from the conglomerates of the Yuryuzan-Sylvino depression of the Pre-Uralian trough; and almost 50% of the area around Bashkiria contains sandstone karst. There are numerous caves in conglomerate and sandstone in the Kama-Middle Volga, the Caucasus, the Lena-Enisei and the Sayan areas, and also in the Ural Mountains (Lyakhnitsky and Khlebalin, 2006). Vdovets (2006) reports numerous 'clastokarst' caves formed by removal of soluble (carbonate) cement from sandstones, the Bolshaya Oreshaya Cave being the best known.

Large numbers of limestone caves, but few caves in sandstone, are known from western Europe and northern America. Mullan (1989) and Self and Mullan (2005) reported 10 small caves in the Fell Sandstone, a quartz-cemented arenite in northern England. These caves, the largest of which is only 9.6 m long, apparently formed by piping through arenized sandstone. As the area was recently glaciated, these caves and nearby karren forms apparently developed rapidly during the Holocene. Caves in marly sandstone in Italy are up to 7.01 km in length (Mocchiutti and Maddaleni, 2005), but dissolution of quartz has not played the major role in the development of these carbonate-rich rocks. A series of short caves, springs and closed depressions have been described from the Precambrian Hickey Sandstone of Minnesota (Shade *et al.*, 2000).

Numerous caves in quartz sandstone occur in Australia, especially in the seasonally dry tropics (Fig. 6.12). They are, however, much smaller than the large caves in South America and Africa, probably because of the greater aridity of Australia. The largest seems to be the Whalemouth Cave, about 20 km west of the Bungle Bungles. It is approximately 220 m long, 120 m deep and is fed from a doline that extends down 40 m below a dry valley now preserved on top of the adjacent cliffs (Jennings, 1983; Grimes, 2007). The exit to the cave is about 60 m high and 45 m wide, dimensions that are partly the result of unloading along

Fig. 6.12 A tunnel cave in northern Australia. Yulirienji Cave near the Roper River is the 'remnant of a former river cave, through ... Upper Proterozoic quartz sandstone' (Jennings, 1983, p. 24). Note the tafoni (windows) in the sides of the tunnel. (Photo: the estate of the late J. N. Jennings)

joints. The cave extends through quartz sandstone that Jennings described as a 'mechanically sound and chemically resistant rock par excellence'. Jennings attributed its development to solutional attack along quartz crystal boundaries followed by removal of the loosened grains. But, as a silica content of only 6 ppm was measured in the stream through the cave, and as annual rainfall is only about 600 mm, Jennings concluded that the origin of the cave goes back to wetter conditions in Tertiary times. Caves and tubes riddle the highly quartzose Kombolgie Sandstone of the Northern Territory, and other caves have been identified near Roper River (Jennings, 1979).

Although the Jurassic sandstones around Carnavon gorge, in central Queensland, contain only one cave of moderate size – the archaeologically famous Kennif Cave, which has about 100 m of passage (Joyce, 1974) – the area is notable for the extremely large number of small tubes that discharge water well into, and in some instances throughout, the dry season (Young and Wray, 2000). There are 1–15 tubes per square metre. Some are 50 cm or more across, but they are typically 1–1.5 cm in diameter, and can be traced for at least 10 m into the sandstone. The internal surfaces of most tubes are case-hardened by secondary silica deposits, suggesting that they formed under phreatic conditions below the water table (Figs 6.13, 6.14).

Fig. 6.13 Cathedral Rock in Carnarvon Range, central Queensland, Australia.
Note the many tubes pitting the face, and runnels indicating solutional weathering

Fig. 6.14 A large tube entrance on Cathedral Rock, Carnarvon Range, has
siliceous case hardening on the wall

The sandstones of southern Australia have numerous small tubes, but few caves or tunnels. One major sandstone cave is known, the 85-m long active stream cave of the Natural Tunnel at Hilltop, south of Sydney, formed through the Hawkesbury Sandstone (Pavey, 1972). In many respects, this cave is similar to sandstone caves reported from Brazil. The wider significance of these tubes and occasional caves is that they occur in a region where there is no evidence of tropical weathering during the Cainozoic, and perhaps even since the Paleozoic (Bird and Chivas, 1995), and thus they provide unequivocal evidence of widespread solutional weathering of sandstone under temperate climates.

Dolines and shafts

In addition to sub-horizontal cave systems, numerous dolines and vertically walled shafts have been described from sandstone and quartzite. The most spectacular are the huge vertically walled shafts (*simas*) in Venezuela, 150 m to nearly 400 m deep and 100–400 m wide (Urbani and Szczerban, 1974; Pouyllau and Seurin, 1985; Mecchia and Piccini, 1999); similar shafts are found in Guyana (Urbani, 1977) and Brazil (Wernick *et al.*, 1977). These shafts are believed to have formed by roof collapse over underground voids enlarged by basal erosion by underground flow. Because large springs spout from cliff walls near the shafts, huge cave systems must exist below them; and exploration of Auyán-*tepui* by Mecchia and Piccini (1999) revealed cave linkages between several of these shafts. For example, a surface stream on the northwestern part Auyán-*tepui*, plunges down a sinkhole in the flat top of the *tepui*, into the extensive Aonda Cave System, and emerges from the huge vertical cliffs of this table mountain. Piccini (1995) proposed an evolutionary model for these *simas*:

- A fracture is enlarged by solution until it reaches an important lithological change.
- In response to this change, and over a very long time, inter-stratal conduits form a drainage network with horizontal water flow allowing the piping of arenized rock and enlargement of the conduit.
- The conduit enlarges laterally along the main axis of the network until its dimension becomes so large as to cause the collapse of the overlying rock, which in turn is mechanically removed through the cave system.
- The cavity so formed enlarges towards the surface until it forms a *sima*.

Vertically walled dolines or *furnas,* some containing deep lakes, occur in Parana State, Brazil (Bret, 1962). The largest of these is circular in plan, 50 m in diameter and 112 m deep, the lower 48 m being filled with water. Collapse *furnas* also occur in sandstones of the Vila Velha State Park, southern Brazil (De Melo *et al.*, 2004). Less spectacular dolines are found in many sandstones which show

Fig. 6.15 The Big Hole, a doline in quartz sandstone south of Sydney, Australia. Ferns colonize the damp base of the hole. (Photo: A. Spate)

solutional features such as caves or runnels – in Africa in Chad (Mainguet, 1972), Niger (Busche and Sponholz, 1992), Transvaal (Martini, 1982) and Zimbabwe–Mozambique (Aucamp and Swart, 1991); in England in the Millstone Grit (Battieau-Queney, 1984); in New Mexico (Wright, 1964; Michael, 1965); and in the Northern Territory of Australia (Twidale, 1987; Nott and Ryan, 1996).

Although the process of subjacent solution of limestone has repeatedly been employed to account for many collapse features in numerous sandstones world-wide, often no evidence for underlying limestones has been found. The majority of these shafts are found to only penetrate quartzites, sandstones and other highly siliceous sedimentary rocks. The Big Hole near Braidwood, about 200 km south of Sydney, is a case in point (Fig. 6.15). This roughly circular shaft in Devonian quartz sandstones and conglomerate is 110 m deep and 30–50 m in diameter, and is located on top of a hill. Jennings (1967, 1983) attributed this shaft to subjacent limestone solution and collapse into the resulting void. But, although limestone is found in several locations within about 10 km, no evidence for limestone has been found in the immediate vicinity or in rubble at the shaft base. Moreover, our SEM examination clearly shows that intense etching and removal of silica in solution has occurred from within the sandstone at the base of the shaft.

Ruiniform landscapes

Corridors and grikes

The widening of joints in sandstones has also produced landform assemblages similar to the grikes and corridors that are commonly found in limestone. While the widening appears to be the result of the dissolving of silica and its removal in solution, and also the removal by running water of the undissolved residuum, the pattern and spacing of fractures in the sandstone is obviously a major constraint on the development of the landforms. The fracture network can be variable. Large fractures (lineaments) extending for several kilometres often exert a major control on streams and escarpments. Smaller fractures (joints) usually intersect, breaking the rock mass into large blocks. Smaller streams may weave along these intersecting fractures. For example, Mainguet (1972) noted in the Hombouri of the Sahara, crevasses a kilometre or so apart, intersecting joints breaking the surface into blocks of about 0.5 ha, and a still finer mesh of joints about 10–20 m apart.

Although many joints in sandstone are only a few metres wide, some fractures are comparable to the giant grikelands of some limestones. Roraima is again the archetypical location. Complex networks of long, narrow, deep joint-controlled slots or crevasses on the summit of Auyán-*tepui* have been known since the pioneer work of Tate (1938), and were superbly displayed in the excellent photographs of George (1989). Piccini (1995) concluded that they are the result of solution progressively widening joints that may have been initially opened by stress release adjacent to the major cliff-lined escarpments. Doerr (1999) described crevasse-like fissures on Kukenan-*tepui* that form large intersecting corridors or labyrinths, and attributed them to solutional weathering of stress-released joint networks.

Extensive complex, grike, corridor and labyrinth networks occur in sandstones in many other places. De Melo *et al.* (2004) describe them from the 'stone city' landscapes in Paleozoic sandstones of the Vila Velha in southern Brazil; and intersecting corridors in central Thailand that were originally believed to be developed in limestone are actually in quartz sandstone (J. Dunkerly, *personal communication*, 1993). There are spectacular corridor and grike networks in the Proterozoic sandstones of northern Australia (see Fig. 6.1). For example, the Ruined City in the Bessie Creek Sandstone in Arnhemland consists of parallel walls and towers separated by networks of 'streets' and open spaces (Jennings, 1983). And numerous similar large corridors and smaller grike networks occur in the Devonian quartz sandstones around Mt Mulligan in northern Queensland (Spate, 1999). The best examples of grike and corridor assemblages in southern Australia are in the Grampians Range of western Victoria, where the weathering

Fig. 6.16 Closely spaced joints have been preferentially eroded to form long corridors in the Grampians, Victoria, Australia

of joints has produced intersecting streets several hundred metres long and tens of metres deep (Fig. 6.16).

Sandstone pavements can exhibit smaller-scale but well-developed grike systems. At Jervis Bay, south of Sydney, highly quartzose marine shore platforms are dissected by rectilinear joint-controlled networks of narrow, deep grikes (see Fig. 5.1). These sandstones show extreme solutional weathering (Wray, 1997b) that has resulted in considerable weakening and enhanced friability of the sandstone. Blocks of sandstone detached along bedding planes, and thus the equivalent of limestone clints are scattered across these platforms. More extensive assemblages occur inland from Jervis Bay, in Monolith Valley.

Although the various examples described here appear to have been formed by the widening of joints by weathering, it is obvious from the earlier discussion of block gliding that many joints are opened by stress release and mass movement. Robinson and Williams (1994) argued that the joints on many sandstone pavements in Europe could not have been opened up by solution, and therefore are not really the equivalent of limestone grikes. They proposed that such joints may have opened as a result of the movement of the sandstone blocks caused by cambering, as for example in the Ardingly Sandstone, which overlies and moves on the Wadhurst Clay. They argued that the fissures at Fountainebleau also might be the result of block movement caused by the loss of unconsolidated sands from underneath the sandstone cap. But they also noted that in many cases, there is no evidence for such movement, and that the joint faces have weathered back, rather than moved apart, and that therefore chemical weathering along closely spaced joints seems to have been important.

Tower assemblages

As joints are deepened and widened, both by solution and erosion, the sandstone between them is shaped into towers or pinnacles of variable shape and size. Where fracturing is intense, sandstone may be progressively developed into a virtual labyrinth of cliff-bounded towers and pinnacles. Mainguet (1972) called this type of landscape 'ruiniform', and gave as a type example the intensely dissected surfaces of the Gara sandstones in Mauritania. Landform assemblages of this type occur extensively in sandstones across the Sahara, through Tassili and the Plateau du Djao (Busche and Erbe, 1987) in the centre, to the Nubian sandstones of Egypt, and thence to the rugged sandstone terrain of the Jordan (Osborn, 1985). Some of these assemblages are truly astonishing, notably the incredible arrays of towers cut in Cambrian sandstone that jut up abruptly from extremely flat erosional surfaces in the *Ruinenlandschaft* of the Tibesti of the central Sahara (Furst, 1966). Fine examples of ruiniform landscapes also occur in the southwestern United States, especially in the Cedar Mesa Sandstone of Canyonlands National Park, Utah. During the last decade or so, there has also been an increasing international awareness that the spectacular tower karst of southeastern China is developed not only in limestone, but also in sandstone. But, yet again, the best known of all ruiniform landscapes is the array of great *tepui* mesas cut in the Roraima Group.

The summits of these almost inaccessible mesas (*tepuis*) in the Roraima are commonly mist-shrouded and experience very high precipitation, receiving an average of 2800–7500 mm annually (Chalcraft and Pye, 1984; Doerr, 1999). As noted previously, much of the runoff finds its way down dolines and shafts into the huge underground cave systems, but some flows directly over the rim, forming spectacular waterfalls that plunge hundreds of metres to the forests below. But the summits are characterized not only by shafts, dolines and grike systems, but even more so by arrays of steep-sided towers. White *et al.* (1966), Urbani (1977), Martini (1982), Chalcraft and Pye (1984), Pouyllau and Seurin (1985), Briceño *et al.* (1990) and Yanes and Briceño (1993) all described large residual quartzite towers and smaller pinnacles on the summits which conform to the definition of tower karst in all aspects; and most of these authors have argued for true karst processes in the formation of these towers.

Piccini (1995) noted two types of rock towers and pinnacles that are the most abundant positive landforms on Auyán-*tepui*. Near the border of the plateau, towers of quadrangular shape from ten to over a hundred metres high are common, and are due to solution–erosion processes along fractures opened by scarp-release stresses. Further back from the plateau rim, the towers are smaller, and are most abundant near the edges of small step-like secondary scarps on the plateau surface. The origin of these towers he believed to be due to solution processes acting along a regular network of two or more joint sets to open fissures, followed by fissure enlargement to leave isolated towers. Ruiniform tower assemblages are also reported in South America from the Vila Velha region of southern Brazil by De Melo and Coimbra (1996). Although solution of silica from the highly quartzose rock is the critical factor in the development of the towers of Roraima, erosion by running water and mass collapse is also very significant. The observations reported by numerous authors (e.g. Doerr, 1999), and the superb photographic record compiled by George (1989), make this clear.

This is true also of sandstone tower assemblages elsewhere. Although smaller than the enormous features of the Roraima, tower assemblages are widespread in northern Australia, being scattered across some 2000 km from northern Queensland to the Kimberley region of Western Australia. Twidale (1956, 1980) reported small towers or 'bee hives', which are about 6 m high and 3 m in diameter, in northern Queensland. Similar small convex forms occur widely in horizontally bedded quartz sandstones, but there are also much larger towers elsewhere in northern Australia.

The classic example of sandstone tower karst in northern Australia is the 'Ruined City' (Boorlungu) of southeastern Arnhemland, which E. Sherbon Hills, when he viewed it from the air, quite understandably considered to be limestone tower karst. This amazing array of towers certainly rivals in appearance the classic examples of limestone tower karst, but it is in the quartzose Bessie Creek Sandstone. Jennings (1983, p. 21) described it as being 'chopped up by meshes of corridors and canyons, and in parts reduced to towers jumping out of the plain' (Fig. 6.17). Percolation of water down and along joints during a long period of sub-aerial weathering has removed much of the quartz cement; and later erosion of this weathered rock, dominantly along the major

Fig. 6.17 Rectilinear meshes of corridors separate towers in the Bessie Creek Sandstone at the Ruined City in Arnhemland. (Photo: the estate of the late J. N. Jennings)

joints, has resulted in the formation of a very striking ruiniform relief, with towers at several levels. Subsurface weathering was certainly significant in the development of this tower landscape. Springs issuing from small tubes along bedding planes, and numerous large closed depressions attest to a perseverance of this underground drainage to the present day. Jennings (1983) insisted that the 'Ruined City' – and presumably the numerous other similar landscapes over much of the Arnhemland Plateau – are attributable to deep solution through the sandstones, and are thus true karst.

The similar ruiniform relief or 'stone forest' in the MacArthur River catchment about 300 km further south, has been eroded from the quartzose Abner Sandstone (Aldrick and Wilson, 1990). It consists of very closely spaced, slender, vertical pinnacles about 20 m high, many of which have height-to-width ratios of about 5:1, and even 10:1 in some cases. Although weathering has penetrated deeply into the sandstone masses of the region, the rectilinear alignment of the narrow gullies between the pinnacles shows that weathering and subsequent erosion was concentrated down vertical joint planes. Cross-bedding on the towers can produce irregular shapes; and a change from intersecting joints to a single dominant joint orientation is reflected topographically in a change from towers to elongated rock ribs.

Fig. 6.18 An isolated tall tower in Litchfield National Park, south of Darwin, Australia represents the penultimate stage in erosion of tower landscapes. The shape of the block is controlled by the variable bedding and strong vertical joints

The form into which the karst develops is constrained not only by the pattern of jointing, but also by the properties of the sandstone. Whereas the Bessie Creek Sandstone has been consumed to the degree that only individual towers, many of which are widely spaced, remain in the Ruined City, far less of the more resistant and less permeable Kombolgie Sandstone has been removed in the development of tower assemblages in northern Arnhemland. Although the numerous caves and tubes in the Kombolgie Sandstone attest to the importance of the subsurface solution of silica from this highly quartzose sandstone, the complex tower terrains east of Jabiru have been formed by a deepening rather than a widening of joints. The corridors between the towers are narrow, and much of the plateau surface remains essentially intact. Nevertheless, tower assemblages even in tough quartz sandstones are eventually consumed. The penultimate stage can be seen along the Table Top Range, in Litchfield National Park, about 100 km south of Darwin. There, isolated towers, still with vertical and joint-bounded sides, rise abruptly from a surface on which scattered low mounds of sandstone blocks attest to the final collapse of towers (Fig. 6.18).

Preferential extension of weathering and erosion along joint systems seems to have played the dominant role in the development of tower assemblages over much of northern Australia, but this was not so in the development of most towers of the Bungle Bungle Range, as was pointed out earlier. On the contrary, many joints there acted as preferential lines of case hardening in the sandstone mass within which silica cement and quartz overgrowths had been extensively etched. Moreover, the multiple levels of towers in the Bungle Bungle Range indicate that renewed incision and the onset of a new phase of denudation occurred before the destruction of the previous land surface was completed.

Quartz sandstone tower karst in Australia is not limited to the tropics. Within both the Sydney Basin and the Grampians, there are large numbers of near-vertically walled towers, as well as many beehive-shaped sandstone turrets locally known as pagodas (Fig. 6.19; see also Fig. 4.5). The summits of plateaus are undulating, but their surface irregularity and degree of dissection increase towards the edges; and in some regions, there is a zone of dissected towers and

Fig. 6.19 Towers in the Grampians, southern Australia, stand in front of the main cliff face. Their surfaces are patterned by erosion along bedding planes and by tessellation

pagodas, below which there are generally continuous cliff lines. Both towers and pagodas are formed by widening of joint networks, but the less steep and strongly stepped slopes of the pagodas are the result of localized weathering and under-cutting of closely spaced bedding planes in the sandstone. Solution of silica plays a critical role in the breakdown of the sandstone. The original network of interlocking quartz overgrowths has been subjected to very intense solutional weathering, leaving a friable, poorly cemented sandstone that remains strong in compression but is weak in shear and tension, facilitating erosion.

Eastern Europe has many 'rock cities', especially near Ardspach and Teplice in the Czech Republic (Migon *et al.*, 2002; Urbanova and Prochazka, 2005) and the Stolwe Mountains of Poland (Latocha and Synowiec, 2002; Alexandrowicz and Urban, 2005). Renewed interest in these rock cities has thrown light on the development of tower assemblages in temperate environments. Cilek (2002) proposed a four-stage development sequence for the sandstone towers of Bohemia (Fig. 6.20). These cities are a maze of narrow gorges or fissures often only 1 m wide that follow the tectonic setting (joints). The vertical rocks, overhanging

Fig. 6.20 Impressive smoothed spires rise above forest in Bohemia. (Photo: J. Dixon)

cliffs, detailed surface morphology and occasional small caves are formed by a succession of phases:

- Firstly, sandstone is submerged below the groundwater table where circulation takes place through the pores and joints.
- After uplift and/or downcutting, the softened sandstone is removed and the vertical fissures are exposed to erosive processes such as freeze/melt during the glacial periods.
- In the subsequent mature phase, uplift erosion has ceased, but exposed surfaces have been modified by weathering processes, especially those associated with capillary water. Cilek considered that biological erosion and salt weathering are the most important erosive processes, and that case hardening by free silica helps surface stability. Exfoliation or desquamation scaling is an important erosive process.
- The continuing action of creep, salt and biological weathering and cambering continues through a senile phase and the ultimate reduction of the towers.

Alexandrowicz (2006) and Urban *et al.* (2006) suggested that 'rock cities' in fine to coarse sandstone and conglomerate in Poland have been shaped by similar processes. The size and shape of the towers depends mainly on lithological features of the sandstones and the density of the jointing. These rocky outcrops have then been exposed by erosion, gravitational processes and weathering; although case hardening, especially of ferruginous materials, has helped to preserve the shape of the towers.

7

Erosional forms

Bare rock surfaces

A characteristic feature of many sandstone terrains is a substantial exposure of rock that is devoid of soil cover. This is not only so in areas of extensive slickrock slopes, but also in areas which are partly forested, or have extensive cover of heath or sedges. For example, the sandstone plateau surfaces of the Sydney Basin are covered by a mosaic of woodland, heath, sedgeland and bare rock; and rocky outcrops and peaty moors form a complex patchwork on the Fontainebleau massif in central France (Thiry and Liron, 2008). The bare areas are not simply surfaces that have been previously soil-covered and then stripped by erosion. In most situations, they seem to be permanently exposed outcrops, albeit often gently sloping and adjacent to similar slopes that are soil-covered and vegetated. Wilkinson *et al.* (2005) found in the Blue Mountains, Australia, the bare rock surfaces weather and erode at comparable rates to the adjacent vegetated areas – more quickly than slopes covered by forest but more slowly than heath-covered areas. Mean erosion rates were estimated to be 13 m/Ma on the bedrock outcrops, 10 m/Ma beneath forest and 17 m/Ma beneath heath. The rock breaks down by granular disintegration, and removal of loosened grains.

Bare rock outcrops are eroded by runnels which cut across them, by basins which are etched down into them and by removal of disaggregated grains or flakes of rock spalling from the surface. They are hollowed out by cavernous weathering, and this causes not only collapse along the sides, but also lowering of the outcrop. Water seeping into the outcrop promotes caverning, undermining the upper surface of the outcrop. Eventually, the often case-hardened roof of the cavern is breached and collapses, and the floor of the cavern becomes the new surface of that part of the outcrop.

Clearly, solutional decay is important on bare rock outcrops, as is the role of lichens and algae growing on the rock. However, because the surface is exposed, the role of thermally induced stress – whether by extreme heat or by cold and frost – is perhaps more important for these outcrops than for other parts of the sandstone terrain.

The role of thermal stress

Disintegration due to frost action is very important in highly porous and fractured rocks, like sandstone, under very cold climates. The breakdown of rock by fatigue, induced by high-frequency cycles of solar heating, also has long been considered to be of importance, especially in arid lands (e.g. Ollier, 1969; Coque, 1977; Smith, 1977). Fires also can cause spalling and rock breakdown (Adamson *et al.*, 1983). Although the phenomenon of thermal shattering is well recognized in a range of environments, the precise nature of the process of shattering is not universally agreed upon.

Yatsu (1988) has taken up early criticism of the concept of thermally induced fatigue in rocks. His scepticism derived mainly from the classic experimental work by Griggs (1936) who showed that, in the absence of water, negligible fracturing occurred during repeated cyclical heating and cooling of granite. Caution is needed in applying results from crystalline rocks to highly porous sedimentary rocks such as sandstones. This is especially so because detailed experiments combining the heating of dry samples and cooling in water (Journaux and Coutard, 1974) have demonstrated that thermally induced fatigue in rocks is greatly promoted by the presence of moisture. The reason for the enhanced breakdown is probably the marked reduction in tensile strengths from dry to wet samples noted previously. Over a temperature range of 10–40 °C, uncompressed water expands by 1.5%; and if the walls of small pore spaces prevent pore water from expanding, it can exert pressures of up to about 40 MPa (Franklin and Young, 2000). High diurnal temperature ranges combined with rain showers or heavy dew would thus seem to be the ideal environmental conditions for this type of breakdown. Nonetheless, the recent review by Turkington and Paradise (2005) indicates that insolation is unlikely to be the primary cause of the rapid breakdown of sandstone surfaces.

Yatsu argued that the main thermal effect on rocks is due to shock rather than fatigue, and considers fire, rather than insolation, to be by far the more important agent. Sandstones are particularly susceptible to thermal shattering triggered by fire because of their high quartz content and because of their relatively low tensile strengths (Fig. 7.1). Quartz expands about four times more than feldspars and about twice as much as hornblende, making it the most critical mineral in rocks undergoing intense heating (Winkler, 1975). Over the temperature range from 50–570 °C, the linear strain along the *c*-axis of a quartz crystal increases from 0.7–1.46% and the volumetric change increases from 0.17–3.76% (Skinner,

Fig. 7.1 Spalling from quartzose sandstone after a bushfire in the Keep River National Park, northern Australia. Broken fragments have fallen onto the charcoal from the fire, and lichens have been burnt off the rock, exposing the fresh light-coloured sandstone on the outcrop

1966). A further rise in temperature causes a spontaneous conversion of low-quartz to high-quartz with an accompanying increase in linear strain to 1.76% and volumetric change to 4.55%; above 580 °C, strain decreases slightly (Skinner, 1966). The volumetric increase when temperatures rise to about 500 °C produces stresses of 100–250 MPa (Winkler, 1975), which far exceed the tensile strengths of most sandstones. A thin, highly heated layer with a steep thermal gradient causes spalling of rock, although the potential for spalling depends also on the thermal expansion and stress–strain characteristics of the rock.

The response of quarztose sandstone to these thermally induced stresses is well illustrated by the field studies in the Sydney Basin by Adamson *et al.* (1983). Their observations show that fire moving rapidly over a rock face will cause only minor spalling or granular disintegration. Where the rock does not flake, the fire chars the algal coating on the rock and weakens the cement to a depth of 1–2 mm, thereby allowing removal of the grains during subsequent rains. Brief, but more intense fires may cause spalling to depths of 2 cm. On one outcrop, about 50% of the surface spalled, producing about 6 kg/m^2 of debris. The greatest effects are on surfaces where logs are left smouldering for days or even weeks. Clearly, the intensity and the duration of the applied heat is the critical factor, but the mode of disintegration also varies with degree of cementation of the grains. SEM analysis showed that in sandstone cemented weakly by clays, the thermally induced fractures pass between the quartz grains; whereas in sandstone strongly cemented by iron, the fractures cut through the quartz grains. The iron-oxide-indurated fragments remain after grains from the shattered sandstone have been washed downslope. Platey iron-cemented fragments comprise about 90% of coarse material in hillside deposits in parts of the Sydney Basin. Fire is thus an important geomorphological agent in well-vegetated sandstone terrain like the Sydney Basin, where, on average, rock faces are likely to be exposed to rapidly passing fires once in 10–20 years, and to be severely affected by heat from burning logs about once every 6000 years (Adamson *et al.*, 1983).

A review of fire-prone regions (Shakesby *et al.*, 2007) has shown that the rate of breakdown of rock surfaces by fire may be as much as one or two magnitudes higher than that caused by frost. This review also emphasized that fire may significantly reduce the stability of soil, and alter its potential for absorbing or repelling water. These changes have implications for infiltration, rain-splash detachment and overland flow.

Hillslopes and valleysides

Below clifflines, on the sides of V-shaped valleys, and on hills that lack prominent caprock, slopes cut in sandstone are often relatively straight and steep (>10°). Benches that follow bedding planes or changes in lithology are common if the sandstone is horizontally bedded or gently dipping. Where rates of weathering exceed those of erosion, sandstones generally are mantled by sandy to loamy soils, which often have quite marked changes in texture down the profile. These soils have variable depth, ranging from several metres on many gentle sloping interfluves or foot slopes, to a few millimetres on steep slopes or on platforms underlain by particularly resistant beds. The commonly encountered downward-fining within soil profiles has been widely attributed to the eluviation or breakdown of clays, but often it may be due either to selective downslope transport or to bioturbation (Paton *et al.*, 1995; Young and Young, 2001). Mass movement, apart from sporadic cliff collapse along the face of benches, may be rare. The short-term monitoring by Humphreys and Mitchell (1983) indicates that soil creep is negligible on the weathered sandstones around Sydney. Similar conclusions were reached by Williams (1973), who estimated that rain wash on sandstones at sites in northern and southern Australia is 5–7 times more important than creep.

In many tropical and arid lands, and to a lesser degree in temperate lands, sandstone interfluves are armoured by duricrust that greatly retards erosion. The flat to gently sloping duricrusted surfaces have thin veneers of sand and fragments of detrital duricrust that generally extend to within a few metres of the edge of the mesa. The concentration of this sediment in small wash terraces or behind shrubs demonstrates the importance of sheet wash. Where the duricrust has been exposed, it is pitted with solution hollows and runnels of varying sizes. This is especially true of lateritic duricrust, but many outcrops of massive silcretes also show evidence of solutional pitting, and quite large dolines have been observed on surfaces with thick lateritic duricrust (Twidale, 1987). Yet the main processes by which duricrust breaks down are by basal sapping, which takes the form of caverns eating back into the indurated mass, and by the cambering and toppling of blocks over clayey pallid zones (see Fig. 2.11). Our observations in the Tertiary sandstones of the Lake Eyre Basin, in central Australia, indicate that the plasticity of the pallid zones is probably the crucial factor in determining which of these two processes dominates at a given site.

The interaction of biological and physical processes

Research on sandstone in the Sydney region has revealed close linkages between biological and physical processes on sandstones. Humphreys and Mitchell (1983) reported average rates of mounding of soil by termites and ants of 570–840 g/m^2/year, by earthworms of about 134 g/m^2/year, and by cicadas up to 20 g/m^2/year. Falling trees in this region contribute about another 20 g/m^2/year, though tree-fall on nearby metasediments was as high as 135 g/m^2/year. However, all of these rates are small when compared to the 4470 g/m^2/year of soil mounded by the foraging of lyrebirds (Adamson *et al.,* 1983; Humphreys and Mitchell, 1983). The mounded material, of whatever origin, is subsequently moved downslope by rain splash and by surface wash. Nonetheless, Humphreys and Mitchell noted that at all of their monitoring sites in this region where valid comparison of denudational processes was possible, the mounding of soil by animals or by tree-fall exceeded rainwash by a factor of between 2 and 30. Bioturbation is approximately equalled by surface wash only after the destruction of litter and undergrowth by severe fires. In semi-arid and arid regions also, termites, ants, scorpions and other invertebrates shift large volumes of soil above the land surface, making it available for erosion (Young and Young, 2001).

Tractive force exerted by surface flow varies not only with the slope angle but also with the depth of overland flow, and this tends to be inversely related to slope angle. Thus rates of erosion will tend to be greater on slopes of inter-mediate declivity rather than on gentle or steep slopes. The transportation of material is also influenced by the micro-topography of the slope. On sandstone covered by eucalypt forest near Sydney, average rates of sediment transport by surface wash over slopes of up to 10° ranged from as low as 4 g/m^2/year to about 32 g/m²/year (Humphreys and Mitchell, 1983). After bush fires have destroyed undergrowth and vegetative litter under the forest canopy, accelerated erosion may result in rates of surface wash of 250–850 g/m²/year (Blong *et al.*, 1982). Williams (1973) reported rates of rainwash of about 67 g/m²/year under savanna woodland at Brock's Creek in northern Australia, and approximately 123 g/m²/year on the heavily forested hills in the headwaters of the Shoalhaven River, south of Sydney. Single storms in the Sydney region can strip 10 mm or more from the soil on bare slopes, removing the clay fraction completely and moving sand downslope until it reaches sites of temporary storage behind micro-topographic barriers (Adamson *et al.*, 1983). Sand collects not only behind stones or logs lying across the slope, but also behind dams of transported organic debris that range in height from 1–5 cm, and range in length from a few centimetres to extremes of 3–4 m on very gentle slopes (Fig. 7.2). Adamson and his colleagues pointed out that, as these dams are also preferential sites for the germination of seed after fires, the sandy deposits readily become stabilized. However in sedgelands, we have observed that the sandy flat areas behind the small dams may remain unvegetated for some months, while rapid regrowth takes place at the wall of the dam. This is apparently due to higher moisture content, to concentration of seeds opened

Fig. 7.2 Burnt twigs and leaves form a litter dam in a recently burnt sedgeland. The low sedges grow rapidly after fire

by fire and then washed downslope, and to continued growth of partially buried sedges.

Shakesby *et al.* (2007) concluded that, because of the lack of cohesion and water repellancy of sandy soils on steep slopes under eucalypt forests in southeastern Australia, intense erosion may occur if heavy rainfall immediately follows a major fire. But they also emphasize that this is not common because of a variety of factors that disrupt or provide sinks for overland flow, and that bind loose soil or trap mobilized sediment. These factors include mats of fine roots, litter dams and the activity of ants and small mammals.

Sediment accumulation in upland valleys and the development of dells

The erosion of hillslopes in headwater catchments may release sandy sediment in quantities which the headwater streams are often not competent to move. Sediment accumulates in the valleys, encouraging the development of swampy conditions by trapping seepage and low-flow runoff (Fig. 7.3). These valleys lack continuous open channels and are usually treeless. Such valleys in Europe were

Fig. 7.3 A sharp boundary separates eucalypt woodland from the sedgeland and heathland of a dell (upland freshwater swamp) on the Woronora Plateau, Sydney Basin. Note the absence of a continuous channel, and the dark seepage line in the valley axis

termed *dellen* by Schmitthenner (1925), who described them as flat, elongated and often branched, hollows of even gradient with slopes merging in a gentle curvature and not interrupted by a valley bottom. Similar features in Africa were referred to by Mackel (1974, 1985) as *dambos*. He described them as shallow linear depressions in the upper end of drainage systems that lack marked channels and are swampy and treeless. Mackel (1985) suggested that the sediments in aggradational *dambos* are infilling former stream channels, and this is indicated by pebbles underlying the clays that typify the *dambos*.

In the temperate humid environment of southeastern Australia, these upland swamps or 'dells' are common on the sandstone plateaus, but the sedimentary infills usually do not occupy formerly open channels (A. R. M. Young, 1986). Rather, they are due to the low permeability of the underlying sandstone, the gentle inclination and low average relief of the plateaus, and their headwater position. Weathering of the bedrock under the sediments takes place, but the bedrock–sediment contact is usually clearly defined. The sediments are partially or fully eroded out from the dells episodically, a fact indicated by the variable stratigraphy and considerable range of basal ages of organic-rich sands (determined by ^{14}C) within several dells on the Woronora and Blue Mountains plateaus near Sydney. Most of these dates fell in the range of 17–4 Ka, with several less than 1 Ka and the great majority being less than about 12 Ka. Mean rates of sedimentation in dells on the Woronora plateau ranged from 0.1–2.8 mm/year. Sometimes a slug of sand from an eroded swamp can become stranded and then colonized by sedges, forming a valley-fill swamp (equivalent to Mackel's 'aggradational dambo').

The presence of the dells on the Woronora plateau is closely related to the regional climate. Where rainfall is highest and evaporation is lowest, the dells are numerous (up to 9% of the catchment area) and they are almost entirely treeless with sedge and heath vegetation. As the climatic environment becomes slightly drier, the dells become less numerous and are also less distinct from the surrounding eucalypt woodland. Open woodland (*Eucalyptus haemostoma*, *E. racemosa*) with a ground cover of sedges and low heath replaces the more closed sedgeland and heathland of the wetter dells, and the depth and organic content of the soils decline. Swampy open woodland in the northern Blue Mountains occupies gently sloping and poorly dissected broad ridges, but the bedrock–sediment interface can be hard to identify. These valleys may be more akin to the degradational *dambos* described by Mackel (1985) than to the aggradational features previously discussed.

Mackel (1974, 1985) viewed most *dambos* as active erosional forms. He argued that subsurface weathering facilitates subsequent stream erosion; channel incision leads to headward retreat that removes sediment, exposes indurated lateritic horizons and creates small ferricrete platforms at former stream bed levels. The *dambo* itself extends by build-up of sediment around the margins and by the resultant paludification that causes leaching of iron from ferricrete and from underlying sandstone. Penck (1924) also emphasized the intensification of subsurface weathering, but considered that the dells eroded and lowered the underlying bedrock by mechanical erosion due to a slowly moving mass of sediment. Schmitthenner (1925) agreed that dells eroded the plateaus by a broad areal action of moving debris and saw them as the major sites of plateau denudation. Raunet (1985) included a *dambo* type in his classification of shallow valleys (*bas-fonds*) of Africa and Madagascar. These *dambos* are found in humid areas such as west Kenya with 1400 mm/year rainfall and are characterized by *in-situ* soil weathering rather than colluvial or alluvial infilling. However in the Sahel–Sudan region where rainfall is only 800 mm/year, he found very similar valleys which have up to 1 m of fine fill.

Indeed it is somewhat difficult to assess the significance of climate in dell formation. Schmitthenner (1925) rejected the idea that the dells of southwestern Germany were ancient valleys or due to climatic change and noted that they could be found also in semi-arid areas such as the plateaus of Tunisian Sahel but not in the true desert areas of the Sahara. Penck observed them also in the high but arid Puna da Atacama in the Argentinian Andes, and as we have noted, they characterize humid regions of both Africa and Australia (Mackel, 1985; A. R. M. Young, 1986). Nevertheless, many European geomorphologists ascribe a role to Pleistocene periglacial activity. Louis and Fischer (1979) stated that cold-period solifluction is of fundamental importance and reject Schmitthenner's contrary statement; while Mainguet (1972) considered shallow upland valleys with only minor channel incision to characterize all sandstone areas which had experienced or now had periglacial climates. In contrast, Thomas and Goudie (1985) commented that *dambos* are found in Guyana as well as Africa and added the term 'tropical' to an otherwise morphological description. Raunet (1985) also implied a climatic control in commenting that the shallow valleys with which he was concerned (including *dambos*) occupied the planation surfaces of the stable tropical shield areas of Africa, Brazil, Ceylon and Australia on which there are deep weathered profiles developed on Precambrian granitic–gneissic rocks. On the grounds of geomorphic form, soil conditions and vegetation structure, there seems no reason to separate the dells and dambos of the various regions. They characterize gently sloping sandstone uplands in climatic regions as diverse as southeastern Australia, southwestern USA, southern Germany and northern Africa; they occur on other plateaux with dominantly siliceous lithologies in many other places. Thus they seem rather to be a response to geomorphic and lithological factors than to be the result of past or present climatic controls.

Erosion along fractures

As noted previously, while solution can play the crucial role in the breakdown of sandstone, and thus the widening and deepening of the fractures that split it, the removal of the weathered debris is largely by the tractive force of water. Streams or waves cause the erosion, but the influence of partings – fractures and bedding planes – is usually obvious. Indeed, the topographic imprint of the erosional enlarging of fractures is probably more obvious in sandstones than in other groups of rocks (Mainguet, 1972). This seems true both on rocky hillsides and in stream channels. The greater the resistance of the rock, the more important is the erosional opening of fractures in the denudation of sandstone. In very strongly cemented rocks, erosion is concentrated almost entirely along bedding and joint fractures (Fig. 7.4); in weaker rocks, fractures may have little perceptible effect on the pattern of erosion.

Fig. 7.4 A *geo* near Scrabster, northeast Scotland. Marine erosion widens the vertical joints, forming open fractures. This one is bridged by joint blocks

Where fracturing is intense, sandstone may be cut into a virtual labyrinth of cliff-bounded blocks or pinnacles, the morphology that Mainguet classified as 'ruiniform'. While the erosion of joints is generally dominant, the erosion of bedding planes is also highly significant. For example, the rock cities of Poland have been attributed to an initial stripping along major joints, with subsequent development controlled largely by bedding and lithological variations (Alexandrowicz and Urban, 2005). Of course, such towers are by no means limited to ruiniform landscapes, for in many instances individual towers or isolated groups of towers have been cut along intersecting joints in otherwise uniform cliffs. Gregory (1950) pointed out that the huge towers of the Watchman and the Great White Throne in Zion Canyon are sculptured blocks of relatively unfractured Navajo Sandstone, between zones of more intense jointing through which deep canyons have been preferentially eroded.

Another notable case in point is the Three Sisters, cut from the Triassic sandstones of the Blue Mountains, west of Sydney. Gerber and Scheidegger (1973) used the example of the Three Sisters to argue that the cutting of towers or bastions of this type is to some extent constrained by local stress distributions, and that the resultant forms tend to be very stable. Be that as it may, the Three Sisters have developed primarily as the result of the widening of major vertical fractures. As these fractures extend down into a prominent lower cliffline, it is clear that they were formed by tectonic stress rather than by stresses generated on the towers themselves (Fig. 7.5). The detailed form of each of the three main towers in this group is also partly controlled by undercutting in several prominent claystone beds and in numerous thin beds of friable sandstone. However, the decrease in height from the inner to the outer of the three main towers, and the remnants of other towers which once extended beyond the main group, show that the morphology of the triangular promontory on which the towers stand is the

Fig. 7.5 Sketch of the joints and bedding planes which control erosion of The Three Sisters towers, Blue Mts, western Sydney Basin

dominant constraint on their stability. The more friable beds, on the middle of the canyon wall, form slopes that are gentler than the vertical cliffs below and the towers above. Retreat of the middle slopes leads to the sapping of the foundations of the towers. Thus, as the promontory becomes narrower towards its apex, the width of the middle slopes and the thickness of the towers decrease. The towers, which are already partly undercut along friable beds high in the section, then collapse.

Jointing exerts considerable control on the patterns of drainage developed on sandstones. One of the best known instances is the striking trellis pattern of drainage in the Navajo Sandstone east of Zion Canyon, where tributaries following one major set of joints are aligned at about 90° to the trunk streams which follow the other main set (Gregory, 1950). Another outstanding instance of trellised drainage, developed along intersecting joints, occurs where the headwaters of the East Alligator River drain the Kombolgie Sandstone of Arnhemland in northern Australia (Bremer, 1980). Yet, the occurrence either of trellis or dominantly linear drainage patterns on sandstones is not invariably the result of control by fractures. For example, the dominantly linear pattern of the headwater tributaries of the Nepean River, which are incised into the Hawkesbury Sandstone, west of Wollongong in southeastern New South Wales, mainly follows the regional dip of the sandstone rather than the prominent fractures in the sandstone. Joint control certainly does occur on some of the straight reaches of these tributaries, but is most prominently displayed in the abrupt bends that interrupt the dominantly linear pattern.

The control exerted by intersecting joints on the incised bends of streams has long been recognized (Bach, 1853, cited by Hettner, 1928), especially on streams crossing sandstones. The gorge which the Katherine River has incised into the Kombolgie Sandstone is a fine example, for the river has little capacity to alter the great blocks of sandstone between the fractures, and it therefore turns abruptly from one fracture to another (Baker and Pickup, 1987) (Fig. 7.6). Mainguet (1972, Plate 64) provided another striking example of abrupt, joint-controlled bends in sandstone from the Konkoure in Guinea. She contended that such abrupt bends, or false meanders (*faux méandres*), are widespread and frequently found in sandstones and are indeed more common in sandstones than in any other rock. In many cases, the bends cut by streams into sandstones are morphologically indistinguishable from meanders developed in alluvium (Dury, 1964, 1966). Yet a strong influence of jointing may still be evident. Young (1977b) has demonstrated that the angular bends, initially cut along intersecting joints in the Nowra Sandstone by the Shoalhaven River and some of its tributaries, were gradually modified into symmetrical meanders that conform closely to the sine-generated curves of minimum work that appear to underlie the development of alluvial meanders.

Fig. 7.6 The Katherine Gorge, Northern Territory, Australia. The gorge changes direction abruptly as it flows down pronounced orthogonal fractures in the Kombolgie Sandstone

Young also demonstrated that the size of incised meanders on these streams is controlled by the interplay of bedrock constraints and the magnitude of discharge.

Canyons

Slot canyons

Where sandstones are not strongly cemented, but are still capable of standing in vertical or even overhanging faces, streams often cut very narrow and deep slots into the upland surface. The extremely narrow chasms with vertical and often overhanging walls that are typical of slot canyons, or slot gorges, are well displayed on the Colorado Plateau, as for example in the Narrows of the Paria River and the North Fork of the Virgin River. The main section of the North Fork of the Virgin River Narrows in Zion National Park is in places only about 10 m wide but about 600 m deep, and some of its smaller tributaries are less than 1 m wide but about 70 m deep (Ives, 1947). Although some undercutting due to seepage occurs on the walls of this canyon, and very large blocks have fallen to its floor, the frequent switching of the overhang from one side of the canyon to the other as the river winds through highly sinuous meanders demonstrates that the canyon is essentially the product of the stream incising downwards into the sandstone. Ives estimated that this slot canyon has a sinuosity (channel length:axial length ratio) of about 3. A highly sinuous slot canyon has also been cut by Muley Twist Creek, which is incised into the Wingate Sandstone on the Waterpocket Fold of the Colorado Plateau, and many dry slot canyons – such as Antelope Canyon – characterize the aptly named Canyonlands.

Holland (1977) attributed the numerous slot canyons in the Blue Mountains near Sydney to the headward retreat of nickpoints along prominent claystone beds in the generally weakly cemented sandstones of the Narrabeen Group (Fig. 7.7). He pointed out that the slots terminate abruptly on vertical cliff faces at the level of the lowest of these claystones, and that minor nickpoints occur within the slots where stream beds cross claystones higher in the sequence. Holland's

Fig. 7.7 A slot canyon – the Grand Canyon –
in the Blue Mts, west of Sydney, Australia

contentions are supported by the absence of slot canyons in the southern part of the Sydney Basin, where sandstones are generally more strongly cemented and are inter-bedded only with thin, discontinuous claystone lenses. In the latter region, shallow upland valleys terminate abruptly in waterfalls at the top of the major cliff-forming sandstones. Observations by Wray of these slot canyons suggest that the strength and resistance of the rock, and particularly the wet shear strength, is important; the Narrabeen Group sandstones of the Blue Mountains have a higher clay content and a lower strength than the Hawkesbury and Nowra Sandstones in the southern part of the Basin. Mainguet (1972, pp. 130–132) also saw the cutting of slot canyons as a response to lithological variability, but gave greater emphasis to its control of seepage from the canyon walls. She argued that slots that widen at the base might be the result of the enlargement of joints, both by deepening of the stream bed above a knickpoint and by the vertical growth of cavities beneath the knickpoint. The narrow upper portion is a product of channel deepening, whereas the wider lower portion is the product of cavity enlargement. Increased lateral erosion by a stream during incision offers an alternate explanation for the basal widening of slots.

Wohl (1993, 1998) has drawn attention to the complex bedforms and to the undulating walls of many slot canyons. In the canyon of Piccaninny Creek, which drains the Bungle Bungles in northwestern Australia, a coarse abrasive bedload and highly turbulent annual floods have carved spectacular bedforms, including pools and longitudinal grooves, into the weak sandstone (Fig. 7.8). The longitudinal grooves seem to be the product of the growth and shedding of vortices, in a manner similar to von Karmen vortex streets (chains of paired, oppositely swirling vortices that form downstream of an obstruction in fluid flow). Undulating walls apparently indicate a regulation of flow energy similar to that produced by undulating bedforms. Carter and Anderson (2006) have also emphasized the importance of undulating walls in slot canyons. Their observations in Wire Pass, where the upper Paria River incises through the Navajo Sandstone in Utah, and of flume studies of similar channel morphology, indicate that flow width/depth ratios influence the distribution of shear stress, and thus determine whether the slot widens its walls or incises its bed. Excellent photographs taken in Fat Misery's Canyon and 'The Cathedral' at Pine Canyon and displayed on the Natural Arch and Bridge Society website illustrate the downward-spiralling vortex shapes of these canyons.

Canyon sapping

Slot canyons are the product of dominantly overland flow that carves into the rock. Where subsurface flow through very permeable sandstones becomes locally

Fig. 7.8 Sinuous grooves and deep potholes have been carved into
the bed of Piccaninny Creek, Bungle Bungle Range, Western Australia

more important than overland flow, the dominant process changes from linear
erosion to sapping, and a morphologically distinct type of canyon develops.
Unlike the generally highly sinuous and narrow slot canyons which often have
overhanging walls, the canyons formed dominantly by sapping of sandstone are
much straighter and wider, and have mainly vertical walls. Overhanging sections
in these canyons are limited to sites of prominent seepage. Moreover, canyons
produced by sapping terminate headwards in broad amphitheatres rather than in
narrow clefts (Fig. 7.9). Surface flow over the headwall of these amphitheatres is
generally much less than the groundwater seeping from it, and in some instances
may be negligible. This type of canyon begins by extending linearly owing to the
collapse of the roof of deep overhanging sections formed where the headwall
intersects major concentrations of seepage. The rate of seepage decreases as the

Fig. 7.9 Sapping at the base of the Navajo Sandstone in Zion National Park, Utah, has left a deep high alcove at the cliff base

canyon head approaches the limits of the groundwater catchment (Fig. 7.10), so that headward extension slows down and the canyon widens because seepage from secondary amphitheatres on the sidewalls becomes more important (Laity and Malin, 1985; Howard and Kochel, 1988; Laity, 1988).

Sapping also seems to produce a distinctive drainage network. Laity (1988) reported from a study of 30 canyon systems extending in this fashion through the Navajo Sandstone of the Escalante River catchment, on the northern side of the Colorado River, that the drainage network is dominated by (Strahler) order 1 and 2 branches; and that in none of the networks is the trunk canyon greater than order 3. Moreover, there is generally a pervasive parallelism of the tributaries, and the networks are often highly asymmetric, with tributaries extending almost exclusively up the dip of the sandstone.

Studies on the Colorado Plateau indicate that the development of canyons by sapping is very much dependent on hydrologic, lithologic and structural constraints (Howard and Kochel, 1988; Laity, 1988). Laity listed these constraints as:

- a permeable aquifer,
- a readily rechargeable groundwater system,
- a free face from which seepage can emerge,
- structural or lithologic inhomogeneity that locally concentrates groundwater,
- and a means of readily removing debris that collapses from the walls of the canyon.

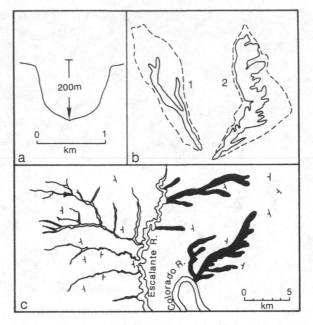

Fig. 7.10 Canyon sapping in the Navajo Sandstone. (a) Cross-section of a canyon formed by sapping at the Inscription House, near Kayenta, Arizona, showing the typically vertical walls and broad floor. (b) Stages in canyon growth: 1 headward extension dominant; 2 canyon widening dominant. (c) Distribution of tapered canyons (*left*) and theatre-headed canyons (*right*) near the Escalante River (after Laity and Malin, 1985). *Shaded areas* represent the canyon floors. Note the relationship between canyon morphology and the dip of the Navajo Sandstone

These constraints explain why the extension of canyons by sapping on the Colorado Plateau is limited largely, though certainly not exclusively, to specific sections of the Navajo Sandstone. The Navajo Sandstone is one of the most permeable of the sandstones of the region. It has a well-developed but not overly intense fracture system and very extensive surface outcrop which permits ready recharge; and it will stand in a free face. Its calcite cement can be leached by seepage, and this produces a generally fine debris that can be readily removed. However, extensive sapping in the Navajo Sandstone is limited mainly to the Escalante River, north of the Glen Canyon section of the Colorado River (Laity and Malin, 1985), and the Navajo Indian Reservation on the southern side of Glen Canyon (Howard and Kochel, 1988). In these two areas, the overlying Carmel Formation, which acts as an aquiclude above the Navajo Sandstone, has been extensively stripped. Moreover, both of these areas have dips of only a few degrees leading down from relatively high recharge areas, and seepage into the rock dominates overland flow where dips are less than 4° (Laity, 1988). Hence, in these areas, the groundwater can be recharged. The importance of local controls on sapping is clearly demonstrated by the contrasts between the canyons east and west of the Escalante River (Laity and Malin, 1985). To the east, where dips are continuous and gentle, and where overburden has been extensively stripped from the Navajo Sandstone, sapping is very well developed, and the canyons are wide with amphitheatre heads (see Fig. 7.10). To the west, where an anticline runs north–south across the drainage lines, and where substantial overburden lies above the Navajo in the headwaters, overland flow is more important, and the canyons are long and narrow.

Groundwater sapping is probably the main cause of the typical box canyon morphology – canyons with a broad floor and near-vertical walls – that is well known from the sandstones of the southwestern USA (Howard and Kochel, 1988). Nonetheless, this morphology is common also in periglacial areas where it is apparently formed by the mechanical weathering effect of an 'ice rind' beneath floodplains, and by lateral migration of braided channels (Büdel, 1982). But none of these mechanisms can account for examples from the semi-arid tropics. The broad, stream-free embayments on the eastern flank of the Bungle Bungle massif (see Fig. 1.5) are apparently carved by sheetwash fed from the steep flanking ridges, for large clumps of spinifex grass retard the concentration of flow across the floor (R. W. Young, 1987).

The amphitheatres developed by sapping of the Navajo Sandstone differ in one essential aspect from the amphitheatres of the southern Sydney Basin. The deeply recessed back walls with extensive overhanging roofs in the Navajo have no real counterparts in the sandstones of the latter region. There most amphitheatres have mainly vertical walls, and the only substantial overhanging sections of amphitheatres are adjacent to waterfalls where outflowing groundwater is concentrated. The contrast is probably the result of the lower permeability and the greater degree of cementing by quartz and iron in the Sydney Basin sandstones.

Channel cutting

Although spectacular, slot canyons and canyons formed by sapping are the exception rather than the rule in sandstone terrain. Most canyons in sandstone are formed dominantly by stream erosion, rather than by sapping, and most have much more open morphologies than do slot canyons. Where major lithological contrasts are present, the walls of canyons are typically stepped. The canyon floors also bear the imprint of structural and lithological constraints. The account of the Grand Canyon of the Colorado by Cunningham and Griba (1973) shows that the details of benching of canyon walls are dependent not only on lithological and structural properties, but also on the distribution of erosive energy expressed in the hierarchy of tributary channels. They made the important observation that the interdependent hierarchies of mass movement and linear erosion control canyon morphology.

Stream erosion in strongly jointed sandstone

In well-cemented and strongly jointed sandstones, stream erosion widening the joints and collapse of joint-bounded blocks from the walls can produce vertical, rather than overhanging walls, even where almost the entire bed of the canyon is occupied by the stream. This is the case for the gorge cut by the Katherine River through the Kombolgie Sandstone in northern Australia (Baker and Pickup, 1987). The floor of the Katherine River gorge has a striking pool-and-riffle morphology developed mainly in bedrock, although some riffles are the result of block fall from the walls of the gorge (Fig. 7.11). The deepest of the pools occur at the abrupt, joint-controlled bends of the gorge, where during flood stages the river locally achieves depths of 45 m. The bedrock bars and major accumulations

Fig. 7.11 A rock bar in the channel of Katherine Gorge. The main channel passes to the left of this gap, but in flood flows, this channel fills, sometimes to the top of the cliffs

of boulders impound the pools even during the long dry season, and thus act in a fashion analogous to the riffles of alluvial streams. One bedrock bar rises up to 4 m above low-water levels, but is incised by a distinct inner channel through which low flows fall about 1.2 m to the next pool. The large pools impounded behind the major riffles are often divided by second-order riffles. As many of the boulders are too large to be moved except by extreme floods, they probably do not contribute to a self-adjusting regime similar to that in alluvial channels, but accumulate because of localized events such as collapse of part of the walls of the canyon.

Baker and Pickup (1987) reported numerous minor erosional features including potholes, polished surfaces and flute marks from rock bars and boulders in the Katherine Gorge. They noted that some of the potholes have spiral grooves that are probably the product of abrasion by sediment-charged water flowing in a vortex. Similarly, Gregory (1950) noted the deepening of potholes in Navajo Sandstone. But in friable sandstones like the Navajo, deepening can be very rapid – measurements over a period of 5 years revealed that of the eight potholes studied, five had deepened by about 7 cm, two by 10 cm and one by 20 cm. Flute marks have been regarded by Allen (1971) as indicative of a variety of mechanisms such as cavitation, fluid stressing, corrosion and abrasion. The fine polishing of bedrock faces in Katherine Gorge is probably due to very high impacting velocities by abrasive particles (Baker and Pickup, 1987).

Young (1977b) observed the quarrying of blocks from the floor of the Kangaroo River, south of Sydney, by the enlargement of joint and bedding planes and by the subsequent lifting of the detached blocks during high flow. The enlargement of bedding planes in sandstone, especially along small channels, may lead to the opening of cavities several metres long under the floor of the channel, and to underground flow that precedes the eventual collapse of the roof of the cavity. Closely bedded sandstone tends to produce slabs that are broad and thin, and thus are particularly vulnerable to lifting during turbulent flow. Young (1977b) also described bedrock undulations on the floor of the channel cut by the Kangaroo River and its tributaries through tuffaceous and quartzose sandstones in the southern Sydney Basin. As the bedrock quarried by these streams breaks down into cobbles that can be transported by the normal flood regime, the spacing of pools and riffles does form a self-adjusting system similar to that of alluvial channels. However, the accumulations of cobbles are superimposed on bedrock undulations of greater magnitude. The channel morphology thus appears to reflect an overall adjustment to a decreased channel-forming discharge in a manner described by Dury (1966) for the Osage-type of underfit stream. The large bedrock undulations were apparently formed during a period of higher discharge, whereas the cobble bars are controlled by contemporary flows. Tinkler (1993) observed that slabs $>1\,m^2$ and about $0.1\,m$ thick were lifted from the floor of Twenty Mile Creek, Ontario, by flows $<0.4\,m$ deep, and we have seen movement of blocks of similar size in similar depths on sandstone channels south of Sydney. Nott (2006) pointed out that as velocity increases, drag forces increase on the upstream side of slabs, and pressure above the slab decreases (the Bernoulli effect), causing the boulder to lift upwards. The slab may consequently be flipped over. However, the force needed to move a slab that is locked against adjacent slabs, and is thus subjected only to lift, is much greater than that for a slab of similar size already detached from the bed.

Waterfalls and nickpoints

The profiles of streams flowing through sandstone are frequently interrupted by nickpoints, rapids or waterfalls, abrupt breaks of slope where the stream bed steepens. The outstanding example is the 972-m drop of Angel Falls from Auyán-*tepui*, the world's highest waterfall. Mainguet (1972) argued that the high frequency of rapids is largely due to undercutting promoted by the differential permeability of successive beds in a sequence of sandstones. Our own observations indicate that the critical factor might be differential strength rather than permeability, or perhaps a combination of both properties. Be that as it may, the profiles of sandstone channels in upland areas are characteristically broken by numerous small steps that are pushed headward until a particular bed pinches out or its morphological effect is overshadowed by another, more prominent bed. Gregory (1950) came to a similar conclusion for the bedrock profiles of streams in the vicinity of Zion Canyon.

In western Africa, Mainguet (1972) observed waterfalls developing at the heads of deep crevasses in sandstones. Evidently streams excavate slot gorges along major fractures, but can erode only very slowly the intact sandstone at the end of the fractures. However, this relationship does not hold at all sites of intense fracturing. R. W. Young (1985) reported counter instances of the suppression of the development of waterfalls by the ready pathways for vertical incision offered by deep crevasses cutting back into cliff faces.

The widespread occurrence of major waterfalls on sandstones is normally attributed to the undercutting of a caprock by the disintegration of a softer rock beneath it. Indeed, as Young (1985) has emphasized, the neglect of research on waterfalls in general can almost certainly be attributed to the common misconception that they can be readily explained in terms of the caprock model outlined in the fine account of Niagara Falls by Gilbert (1896). Yet almost 70 years have passed since von Engeln (1940) demonstrated that the caprock waterfall was just one of several types of falls, and that even this type was not adequately understood. Certainly, many waterfalls in sandstone can be explained in terms of undercutting of a resistant caprock. But von Engeln (1940, 1957) showed from closely bedded sediments on Cascadilla Creek, in the state of New York, that a waterfall could be initiated, and even accentuated or enhanced, when variations in structural resistance are not present. The examples of waterfalls without under-cutting listed by von Engeln were added to by Young (1985), and many of these waterfalls are cut in sandstones. Instead of being undercut, many waterfalls in sandstones are buttressed at the base, and some even lack plunge pools. Three examples that plunge over the Hawkesbury Sandstone, south of Sydney, illustrate this point. Carrington Falls has a slightly buttressed face that terminates in a deep pool. Fitzroy Falls has a vertical face that ends in a pile of boulders rather than a pool (Fig. 7.12). And although there is undercutting at Belmore Falls, it does not

Fig. 7.12 Fitzroy Falls, Shoalhaven River catchment, Sydney Basin. These falls are cut in almost horizontally bedded sandstone but are not undercut and there is no plunge pool

occur on the face of the falls, but rather where seepage flows from the side of the amphitheatre. These examples can be matched by others from quite different settings. For instance, streams flowing over the Kombolgie Sandstone on the Arnhemland escarpment of northern Australia tumble over a vertical face at Jim Jim Falls, and a strongly buttressed face at Twin Falls.

The basic requirement for the development of waterfalls appears to be the outcrop of rock that, because of its lithological and structural properties, will stand in a steep face. Basal undercutting may certainly promote the development of a steep face, but it is not an essential requirement. In short, the discussion in earlier chapters of conditions giving rise to cliffs and amphitheatres can readily be applied to waterfalls (Young, 1985). The sliding or toppling of blocks, and the brittle failure of intact slabs of sandstone, will be promoted by the shear stress of flood discharges passing over the top of the falls. High porewater pressures created by the very steep hydraulic gradients generally encountered by the abrupt drop in water tables below waterfalls will also promote mass failure. Seepage through sandstone, together with spray blowing back from the falls onto the adjacent rock face, will promote biological and chemical weathering. At many sites, undercutting along the sidewall of the gorge, rather than on the face of the falls, shows the importance of seepage. Water seeping into the rock from above the water table emerges along bedding planes or other discontinuities. That it has reacted with the rock through which it has passed is shown by the frequent tufas that develop near the seepage, with the minerals forming the tufas being dependent on the cement of the sandstone.

The specific characteristics of flow over the falls are also important (Fig. 7.13). Wave-like phenomena, associated with pulsating noise of low frequency or with irregular explosive effects, that are known to cause decay by fatigue on the crests of dams (Thomas, 1976), are also likely to cause similar breakdown of sandstone at the top of waterfalls. The energy generated by the falling water, even in the absence of a debris load, can be sufficient to excavate a plunge pool at the foot of waterfalls; the critical depth to which a pool can be excavated is dependent on the height of the fall and the discharge per unit width of the channel. Absence of a pool presumably is due to either a greater resistance of bedrock or the relatively recent retreat of the waterfall. And, as can be deduced from the analysis of form in Chapter 3, the stability of the waterfall will also be dependent on the distribution of stress associated with the shape of the waterfall, both in section and in plan.

Cosmogenic dating of sandstone surfaces in the Grabens region of Canyonlands National Park, on the Colorado Plateau, indicates that small streams incise into bedrock mainly by knickpoint retreat (Phillips *et al.*, 2004). Whereas the knickpoints retreat at rates of 8–10 m/Ka, there is little erosion in the channels

Fig. 7.13 Huge volumes of water cascade over the Shizhangdong waterfall, Chishui, southern China

upstream from them. In southeastern Australia, the dominance of knickpoint retreat over lowering of the bed is also indicated by the widespread occurrence of ferruginous coatings on channel floors and in potholes along many streams flowing on sandstone. Our observations suggest that potholes are largest and most numerous at or close to the top of knickpoints, as are small natural bridges across stream beds (Fig. 7.14). In some cases, potholes act as sumps, and water flowing into them only re-emerges on or at the base of the face of the knickpoint downstream. This indicates an important role for solution, with water seeping through the rock eroding the potholes and widening the bedding planes. The role of solutional erosion in stream bed lowering of sandstone terrain has been almost ignored, but is clearly important.

Stream channel profiles (thalwegs)

Sandstones can exert a major influence not only on the irregularity of stream profiles, but also on the average gradient. In the Appalachian Mountains, at comparable distances from drainage divides, streams crossing sandstones generally have steeper gradients than nearby streams crossing shales or carbonate rocks (Hack, 1957; Brush, 1961). The reason for the greater declivity was attributed to the coarser load that must be moved by streams draining sandstones. However, comparison with measurements of gradients of streams

Fig. 7.14 Water flows under a natural bridge cut in Hawkesbury Sandstone, Sydney Basin. Note the large potholes along and near the bridge

draining sandstones in parts of Australia prompts caution against indiscriminate use of these findings from the Appalachians (Fig. 7.15). For example, the MacDonald River, northwest of Sydney, has steep headwater gradients similar to those on Appalachian streams, but has much lower gradients in its middle reaches where the calibre of its load decreases to a dominantly sand-sized material. On the other hand, the Clyde River which drains the southern extremity of the Sydney Basin, has low gradients in its headwater reaches on an undulating upland surface, then increases its gradient very markedly as it flows through a deep gorge into which boulder-sized material is fed. Streams draining the sandstones of the Kimberley region of northwestern Australia also generally have significantly lower gradients than streams of comparable lengths in the Appalachians, except in the headwater reaches for which there is overlap of data. The reason for the difference in average gradients between the two regions may lie in the attrition of bedload as it is transported away from the main source areas of the Kimberley streams. Whereas the Appalachian streams flow in valleys bounded by ridges that shed coarse debris, the Kimberley streams flow from steep headwater regions out across very broad, undulating platforms on which the only sources of coarse debris are in the floor and walls of the channel itself.

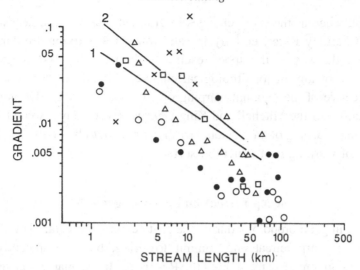

Fig. 7.15 Gradient vs stream length for some Australian streams cutting through sandstones: x Clyde; o King George; △ Prince Regent; □ MacDonald; • Mitchell. Trend lines for eastern USA streams in sandstone are shown: 1 Pennsylvania (Brush, 1961); 2 Shenandoah Valley (Hack, 1957)

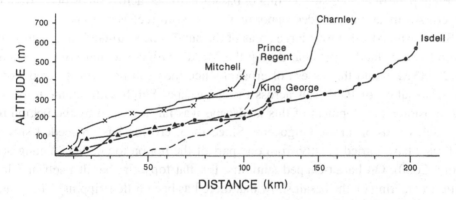

Fig. 7.16 Profiles of streams cut in sandstones of the Kimberley plateau, northern Australia. Note the long gentle central reaches and the abrupt descent over the edge of the plateau

Another reason for the low gradients of the Kimberley streams may lie in local structural control on thalwegs, and in the preservation of flatter sections that originated during earlier cycles of erosion (Fig. 7.16). The long, flat central section of the Charnley River extends across the contact between sandstone and volcanic outcrops, and the very abrupt breaks closer to the headwaters are entirely within sandstones. Several changes in outcrop from sandstone to volcanics along the lower section of the nearby Isdell River seem to have had no major effect on the gradient, whereas the sharp changes in gradient closer to the headwaters are, as on the Charnley River, entirely in sandstone. The striking similarity of the breaks in the thalweg on the upper reaches of the Charnley and the Isdell, both well upstream of any major lithological change, although at different altitudes, seems indicative of the regional warping of erosional surfaces. The long, gently sloping sections of the Mitchell and King George Rivers may also be the product of the regional warping of an erosional surface or, as both lie on the King Leopold Sandstone, of warping of a structural surface.

Scarp retreat and canyon growth

Much debate in geomorphology during the past century was centred on the relative importance of scarp retreat and summit lowering, but until recently far less attention was given to canyon growth. Resolving these matters is particularly significant for the understanding of the long-term development of sandstone terrain, because of the role of sandstone as a caprock and cliff-former. But a perennial difficulty has been that of reliably reconstructing sequential phases in landscape development and of obtaining quantitative evidence of long-term rates of change in different parts of a landscape. Dating of basalts extruded across sandstone terrain in eastern Australia has provided opportunities for such reconstruction.

Basalt was extruded over large areas of the sandstone scarplands and valleys of central Queensland, Australia, during the Middle Oligocene and Early Miocene (c. 20–35 Ma), and the pattern of erosion since then can be reconstructed using the distribution of the basalt across the landscape. Sub-basaltic contours show that erosional development of this landscape was far advanced by the beginning of basalt extrusion in the Oligocene. Since then, erosion has proceeded slowly and rates have varied greatly from one part of the region to another (Young and Wray, 2000). On basalt-capped summits, the flat tops, the basalt itself and lateritic weathering on the basalt show that there has been little stripping. The strata, including the cliff-forming Precipice Sandstone, dip southwards. Canyons have cut 30–60 km back into the basalt-blanketed escarpment on the northern side of the Carnarvon Range, at rates of 1–2 km/Ma. Where the canyons have intersected the underlying shales, they are 6–7 km wide; but where they are cut only in sandstone, they are 3–4 km wide. The rates of valley widening (measured from the edge of basalts in the valley floors) are thus 150–250 m/Ma, a fraction of the rate of headward extension. The rates of scarp retreat are even lower – only 18–130 m/Ma. In this region, headward canyon extension has been at least an order of magnitude faster than either valley widening or scarp retreat, and summit lowering has been minimal, over the past 20–35 million years.

The Precipice Sandstone is a major aquifer, so seepage is important along the escarpment face and on the dipslopes. It emerges as point sources from an integrated network of solutional tubes, initiating linear erosion on the dipslopes. Where the south-flowing tributaries breach the sandstone, they can cut headward along the underlying shales and have breached the escarpment. However, the mean rate of incision along the dipslope streams (10 m/Ma) is comparable to that along the canyons cutting through the escarpment (13 m/Ma) (Young and Wray, 2000). Slickrock slopes on the dipslopes attest to active granular disintegration, and the heights of towers and pinnacles suggest at least 50 m of lowering there.

The dominance of canyon growth over scarp retreat has also been demonstrated from dated basalt in the sandstone terrain of the southern part of the Sydney Basin (Young, 1983c; Young and MacDougall, 1985; Nott *et al.*, 1996). The 600-m high coastal escarpment has retreated at a maximum rate of about 170 m/Ma, while the nearby canyons have been extending headward at about 1.5–2.5 km/Ma. The canyons have deepened at a maximum rate of 13 m/Ma, but in their lower reaches, have deepened at only about 1 m/Ma. Rates of channel deepening on the surface of the adjacent plateau are only about 1 m/Ma, and in some places much less than that. The retreat of canyon walls has proceeded at about 12–25 m/Ma where sandstone is underlain by softer rock, and at about 10 m/Ma where the walls are cut entirely in sandstone. The dominance of canyon growth as compared with scarp retreat is a consequence of the marked concentration of erosive energy in stream channels. A sandstone plateau is consumed more by active headward erosion by streams carving back into it, than by retreat of a long escarpment.

The importance of erosion by streams draining down dipslopes was noted also in a study by Schmidt (1994) of cuesta scarps on the Colorado Plateau. He observed that, whereas scarps cut across the front of dipping sandstones tend to be straight, scarps on the backslopes of the dipping sandstones tend to be much more deeply embayed. Here too, erosion by streams, rather than scarp retreat, is the dominant process.

8

Climatic zonation of sandstone terrain

Lithological and structural constraints on denudation like those considered in the previous chapters have been, by and large, relegated to a secondary role in the shaping of landforms. Their importance is admitted in explaining localized variations in morphology, but generally denied for variations at the regional or subcontinental scale. Rather, it is climate that is given pride of place as the determinant of large-scale variations in style of denudation and morphology. This viewpoint has been elaborated in great detail in the handbooks on climatic geomorphology, especially that by Büdel (1982); more restrained statements have been presented by Tricart and Cailleux (1972) and by Louis and Fischer (1979). In brief, the concept defines morphoclimatic zones, acknowledging the role of past climates, of tectonics and of structure/lithology, but asserting the dominance of climate-controlled processes in shaping the landform assemblages characteristic of the zones:

- Humid tropical zones dominated by chemical and biological weathering, with mechanical weathering and abrasion minimal. Slopes tend to be convex, and mantled by deep weathered regolith. Vertical clifflines are uncommon.
- Arid zones where mechanical weathering (e.g. thermal shattering, wind abrasion) is very much more important than chemical or biological weathering. Prominent cliffs heading pediments that rise from wide plains are typical, as are both dunes and aeolian landforms shaped by wind abrasion. Slope angles change abruptly and slopes, except on crystalline massifs, tend to be straight. Structural constraints are clearly expressed.
- Temperate zones where chemical and biological weathering, though not as intense as in the tropics, nevertheless gives rise to slopes that are generally much less angular than those of arid lands.
- Glacial/periglacial zones dominated by glacial activity and/or frost-related processes. These have the forms due to glacial carving (such as U-shaped valleys) and steep hillslopes covered by frost-shattered debris or solifluction deposits.

The apparent lithological homogeneity of sandstones has made them seem ideal material for demonstrating the supposed dominance of climatic influences (Mainguet, 1972; Tricart and Cailleux, 1972; Cilek *et al.*, 2008). Paradoxically, however, it was in the pioneer work on sandstones by Hettner (1903, 1928), the great advocate of comparative geographical method, that the difficulties of assessing the degree to which climate may have controlled the shaping of landscapes were first recognized. Surprisingly, the objections raised by Hettner have been ignored by modern advocates of climatic geomorphology.

The striking similarities between the cliffed valleys cut in the sandstones of Saxony, and the cliffed canyons of the Colorado Plateau and the wadis of North Africa, gave rise to a highly significant question:

A desert landscape, or should we say more accurately a landscape with desert forms lies in the middle of Germany! Have we therefore once had a desert climate which made these forms?

(Hettner, 1903, p. 609; translated by R. W. Young)

Whereas many of his contemporaries readily answered in the affirmative, Hettner considered that the morphological similarities might be mainly the product of lithology rather than climate.

In attributing the striking rock forms of the Quadersandstein of Saxony or Silesia, or the Bunter Sandstone of the Palatinate to former desert climate, too little account was taken of rock composition, and of the fact that a particular climate and a particular rock composition can simulate one another in their effects, that a porous rock is as dry as a rock in a desert. It is inconsistent of Walther, following what he rather unfortunately termed the 'ontological method', to refuse to allow that a desert once had a moister climate and then with other researchers who follow him, to assume unhesitatingly that Germany had a desert climate even in the recent past.

(Hettner, 1928, p. 13)

Hettner's criticism is not just of historical interest. Not only does it emphasize the complexity of the problem of climatic influence, but it highlights the inherent weakness of using type examples – which has been the dominant methodology employed in tackling this question. His criticism requires us to ask seriously whether the similarities observed in sandstone terrain of, for example, Saxony and the Colorado Plateau might be the result of lithological and structural constraints that over-ride the effects of different climates. It also forces us to assess whether the examples we are using really are representative of the climatic zones in which they are found. This is not simply a matter of trying to decipher the possible imprint on the one terrain of successive types of climate. Can we really achieve much if we assume that climate is dominant, and then attempt to demonstrate that dominance by selecting one example from each of four or five climatic zones? Can we really talk about variations between climatic zones while ignoring the variations within particular zones? Yet this is what has been done in most attempts to assess climatic influence on sandstone terrain, and, for that matter, on terrain cut from other rocks.

The difficulty of selecting a type example for a particular climatic zone is well illustrated in the area in which Hettner developed his critique. The sandstone landscape of Saxony is dominated by cliff-lined mesas, such as Lilienstein; and just across the Czech border the Elbe River flows through a 300-m deep, cliff-lined gorge. However, elsewhere in the Bohemian borderland, erosion of intensely jointed sandstone has produced spectacular arrays of towers in 'rock cities', such as the Bohemian Paradise (Mikulas *et al.*, 2008a). The global distribution of similar arrays of sandstone towers suggests that they are more commonly found in humid than in arid lands, yet the cliff-lined mesas are considered typical of arid regions. But clearly both the current climate and the climatic history of Bohemia and Saxony are essentially identical.

The problem of selecting and interpreting type examples is made all the more complex by the major changes of climate during the Cainozoic. This is especially so in the now cool temperate lands of northern Europe, where those changes ranged the full gamut from glacial to sub-tropical, and perhaps even to tropical. While the effects of former cold climates can be determined by the recognition of undoubted fossil periglacial features, some of the supposed indicators of former hot climates now seem dubious. For example, Büdel (1980) argued that the presence of kaolinitic weathering profiles, and especially of bauxitic profiles in Germany, proved the presence of a seasonally wet, fully tropical climate in the Early to Mid-Pliocene. But Bird and Chivas (1988) have demonstrated from oxygen isotope analysis that the very thick kaolinitic profiles of Permian age in eastern Australia formed under polar to sub-polar temperatures. They cited experimental work showing that the primary requirement for the formation of kaolin is a solution that is only slightly supersaturated with respect to this mineral, and that kaolinite can form under temperatures at least as low as 5 °C. They subsequently showed (Bird and Chivas, 1993) that even the much younger kaolinitic weathering profiles over much of Australia formed under cool temperate to cold climates, rather than the tropical climates previously invoked. Indeed, the temperature under which the Late Cretaceous to Early Tertiary profiles formed was 5 °C cooler than now, and for the Mesozoic profiles it was 10–17 °C cooler than now. Furthermore, Taylor *et al.* (1992) demonstrated that extensive lateritic and bauxitic profiles in southeastern Australia formed under cool wet climates during Eocene times. Clearly, the question of intensity versus duration of weathering arises. Features that may form rapidly under intense tropical weathering regimes can also form by prolonged weathering under cooler climates.

The comparative study which follows is based on our own field observations and on published accounts of sandstone terrain that have enough detail to allow comparison. We consider examples from broad climatic zones, and to simplify the comparison, all of our examples are taken from gently dipping strata. Where possible, we have included case studies used by Tricart and Cailleux (1972) as typifying particular morphoclimatic zones. Our concern is primarily with morphology, rather than processes. In previous chapters we have considered individual processes and their relationship to the varied characteristics of sandstones, and much of what follows is in the light of those considerations.

The tropics

Humid tropics

We compare here three major areas of sandstone topography – the Fouta Djallon of Guinea, the Fourambala of the Central African Republic and the Roraima of Venezuela. In all three cases, the present-day climate is humid, with average annual rainfall in excess of 1500 mm, but with a short dry season.

The Fouta Djallon consists of stepped plateaus, ranging in elevation from 700–1500 m. These plateaus have been carved from a thick sequence of Cambro-Ordovician and Devonian sediments which are composed almost exclusively of sandstones and shales, and which contain several very resistant, well-cemented beds 50–100 m thick. Tricart and Cailleux (1972) divided the region into three main zones:

- The central zone consists of a broad plateau, with an undulating surface of shallow basins and convex hills, in which incision by streams has been negligible.
- The central plateau is flanked by a zone of broad, unevenly dissected benches into which broad embayments have been cut.
- This zone, in turn, gives way to a marginal area that is dominated by cliffed buttes and mesas, which rise abruptly from the surrounding lowland cut across a crystalline basement.

Tricart and Cailleux emphasized not only that abrupt escarpments are common, but that escarpments have remained steep, even though they have retreated over considerable distances.

The apparent dominance of scarp retreat over stream erosion is illustrated further by the profiles of rivers draining the Fouta Djallon. Tricart and Cailleux observed that these profiles are very irregular, typically being comprised of gently sloping reaches separated by waterfalls. Even on the margins of the uplands, where abundant water drops into the surrounding embayments, stream incision has been negligible. For example, the stream that flows over the Souma Falls, near Kindia, has excavated a narrow zone of fractures, without otherwise indenting the rim of the upland (Tricart and Cailleux, 1972, Fig. 1.4). The absence of pebbles, the negligible amounts of silt and the small amounts of sand in channels draining the uplands are interpreted by Tricart and Cailleux as sure signs of the weakness of mechanical abrasion, and thus of the inability of such streams to erode into solid rock. Here, as in many other studies of tropical streams (e.g. Louis, 1964; Büdel, 1982) it is the apparent dominance of biochemical weathering under humid tropical climates – which produces solutes and very fine particles, rather than coarse debris – that is seen by Tricart and Cailleux as the ultimate control on denudation.

Mainguet (1972) gave an alternative interpretation of stream action on sandstones in the humid tropics. She argued that these streams have very irregular profiles not because they erode feebly, but because they almost exclusively excavate along fractures exposed in relatively fresh rock from which weathered crusts have been stripped. Whereas sections of channels lying on exposed fractures may become deeply incised and also cut headwards, reaches above the outcrop of the fracture are only slightly incised because of the greater resistance of the intact rock and because of the subsurface flow of water from surrounding hillsides to adjacent reaches incised below knickpoints. Her argument is supported by the abundant evidence of scouring by sediment-free water below dams. Indeed, the once-accepted notion that stream flow has little erosive power unless it is carrying a coarse load, is at odds with a great deal of field and laboratory evidence. Mainguet also suggested that the supposedly greater irregularity of sandstone channels in the humid tropics compared with those in temperate lands may be less distinct than has been widely believed. We will return to this suggestion later.

Mainguet's studies in the Central African Republic revealed a sandstone terrain with many similarities to those in the Fouta Djallon. In the Fourambala, uplands on weathered sandstones with thin lateritic crusts have broad, undulating surfaces and convex summits, the monotony of which is broken only by concave erosive sections of headwater valleys and by small stepped slopes picked out along structural planes. Moreover, on the margins of the uplands, many buttes and mesas have maintained steep slopes while undergoing substantial lateral retreat. Nonetheless, the bare rock outcrops observed by Mainguet display a variety of forms, such as concave slopes and convex to ovoid summits, and also many smaller features due to solutional weathering of the sandstone.

The complexity of sandstone terrain in the humid tropics becomes strikingly clear when the comparison is extended to include the Roraima (Fig. 8.1). This most spectacular of landscapes is dominated by the colossal ramparts of mesas (*tepuis*), some of which rise abruptly over 1000 m from the surrounding lowlands. Numerous waterfalls tumble over the cliffs, but these waterfalls are by no means the expression of the feeble erosive activity of tropical streams (cf. Tricart and Cailleux), for several enormously deep canyons have cut back into the very centre

Fig. 8.1 Spectacular cliffs tower above forested slopes and debris-scattered ridges in the Roraima of Venezuela. Waterfalls tumble from clefts in the plateau edge. (Photo: S. Doerr)

of the mesas, apparently along multiple joint planes. Furthermore, numerous hanging valleys spilling into the waterfalls have been incised into the summits of the mesas. And, as shown in the previous chapter, the number, complexity and dimensions of solutional features developed in the highly quartzose rocks of the Roraima far exceed those reported from tropical Africa.

Briceño and Schubert (1990) have attempted to explain the occurrence of massive clifflines under the humid climate of the Roraima by attributing them to arid conditions during Pleistocene glacial periods. In support of this argument, they compared the Roraima with the cliffed landscape of the arid southwest of the United States. The similarity is striking, but we suggest that it is mainly due to lithological and structural constraints, rather than to climatic control. The massively bedded and jointed quartzites and sandstones of the Roraima must have very high mass strength ratings and can therefore be expected to form extremely steep slopes. Moreover, appeals to arid climates to explain these cliffs are difficult to reconcile with the abundant and very large karst features which are found not only on the summits of the *tepuis*, but which are also prominent on the faces of the cliffs themselves. These impressive and widespread solutional features would not be expected within a morphoclimatic zone characterized by minimal chemical weathering.

The very considerable contrasts between these three regions highlight the importance of assessing the degree of variation within the humid tropics before attempting to compare them to the landforms of other climatic zones. Other examples emphasize this point. Supposedly, valleys on tropical shields are typically shallow; yet arenites and greywackes in New Guinea are intensely and deeply dissected (Löffler, 1977) and Eocene sandstones in Borneo have scarps and deep, cliff-lined canyons (Besler, 1985). To select a type example assumed to be characteristic of the zone is a far more difficult task than it may have seemed at first. Which is typical of the humid tropics – Fouta Djallon or Roraima? This question is not to be dismissed as just a matter of recognizing the climatic imprint on variable lithology. The prominence of the Roraima cliffs may well be due to the strength of their quartzites, but then why should karst be so much more prominent in these chemically resistant rocks, rather than in the weaker sandstones of the Fouta Djallon? Even selecting common, climatically diagnostic facets from among these three landscapes is no simple task. On the contrary, the attempt to do so raises a fundamental methodological problem. That all three of the landscapes described here have steep rocky slopes and cliffs is at odds not just with the Davisian model of landform evolution, as Tricart and Cailleux were at pains to emphasize; it is also at odds with the tenet of climatic geomorphology which asserts that slopes wane under humid climates.

Seasonally dry tropics

The seasonally dry tropics are widely considered to be the environment *par excellence* for the development of lateritic crusts, and laterite certainly dominates many sandstone landscapes in this climatic zone. Mainguet (1972) described such a landscape developed on the feldspathic Mouka-Ouadda sandstones in the N'Dele area of the Central African Republic. In this region, the weathered mantle masks bedding and jointing, except along the main streams, giving low and very even slopes which are either rectilinear or weakly convex. The convexity of slopes, especially along interfluves, produces a gentle doming that Mainguet likened to a cupola. Because much of the rainfall seeps underground, the valley sides are virtually devoid of minor channels, and there is little movement of clastic debris downslope. Mainguet also described the manner in which surfaces with thick weathering mantles are eventually dissected. Her account of the dissection of such an area in the Carnot region of central Africa is of particular interest, for she emphasized dominance of polyconvex ridges produced by the progressive lowering of the landscape, rather than the parallel retreat of slopes emphasized in many accounts of the seasonally arid tropics.

Fig. 8.2 The Arnhemland escarpment, northern Australia, cut in Proterozoic Kombolgie Sandstone

Where the weathered mantle is thin or absent, structural and lithological constraints are prominent. This is so over enormous areas of sandstone in northern Australia which have never been blanketed by lateritic mantles (R. W. Young, 1987), or from which thin mantles have been largely stripped. Mesozoic sandstones around Mount Mulligan, in northern Queensland, have been carved into cliff-lined mesas. Once again, rounded forms comprise a small part of a landscape dominated by vertical, joint-bounded rock faces, and by stratigraphically controlled benches. Even more spectacular cliffs are cut in the highly quartzose sandstones of the Arnhemland escarpment (Fig. 8.2) and the Cockburn Range (see Fig. 1.2). In central Madagascar, a superb ruiniform landscape has been carved into the Permian marine sandstones of the Isalo massif (Swierkosz, 2002). It consists of a complex array of towers and turrets, cliff-lined mesas, canyons and arches. These examples contrast in striking fashion to the subdued, gently convex forms of Carnot and Moukka-Ouadda, thereby emphasizing the variety of sandstone landforms found in the seasonally dry tropics.

Karst features add to the variety. Extensive cave and tube systems are developed behind the cliffed escarpments of the ruiniform landscapes in the quartzites and sandstones of central Queensland, the Kimberleys and Arnhemland. Yet structural and lithological effects are again evident, especially in the tower landscapes described in Chapter 6. To these Australian examples must be added the many karst features reported by Mainguet (1972) from the seasonally dry tropics of Africa.

Of course, the question of climatic change cannot be ignored, especially in regard to the extensive leaching of silica in the formation of the karst. Nonetheless, there is no good reason to believe that the karst of the seasonally dry tropics is just a relic of previously more humid climates. Water flows or seeps through the karst in northern Australia during the long wet season; and the silica concentrations in streams draining the karst of these areas are no less than in streams draining the huge karst features of the Roraima. The array of sandstone landforms indigenous to the seasonally dry tropics is thus at least as complex as that of the truly humid tropics. Similar forms can be found in both of these major climatic subdivisions of the tropics.

Arid and semi-arid lands

The arid lands and their semi-arid margins have long been regarded as the realm most conducive to the development of bold cliffs, sharp breaks of slope, and the topographic expression of structural lithological constraints. Numerous examples have been used to support this claim. The Adrar of Mauritania is a case in point. Tricart and Cailleux (1972) pointed out that the major cuestas developed on thick sandstones and dolomites in that region have sheer escarpments that retreat mainly due to massive landslides and the subsequent washing of debris from foot slopes. They also emphasized the strong structural constraints on the network of streams draining the cuestas. Giant ramparts occur in other parts of the Sahara, especially in the Borkou and Aguer Tai in Tibesti (Mainguet, 1972) and Tassili (Beuf *et al.*, 1971), where sandstones cap weaker rocks. Where jointing is very strong, as near Kiffa in Mauritania, cliffs combine with angular blocks and towers to give the distinct ruiniform landscape (Mainguet, 1972). Striking instances can be cited also from the giant stairway of cliffs and benches leading up from the great unconformity in the Grand Canyon of Colorado to the summit of Zion Canyon. But perhaps the most well known of all examples are the cliff-lined mesas and buttes which occur to the south of the Colorado canyons, notably in Monument Valley (Fig. 8.3). From examples such as these has come the popular climatic explanation of the dominance of cliffs and a general angularity of topography in the sandstones of arid and semi-arid lands. The explanation is that the low rainfall minimizes weathering but, because of a thin vegetative cover, maximizes erosion. Although plausible, this apparently straightforward linking of climate and landform faces several major difficulties.

Firstly, sandstones within even a single arid landscape have diverse morphologies related to differing lithologies. For example, Roland (1973) has shown that the Permo-Carboniferous sandstones on the northern flanks of the

Fig. 8.3 Vertical cliffs and towers in the De Chelley Sandstone, Monument Valley, Arizona, USA. This landscape has repeatedly been cited as the type example of sandstone terrain developed under arid climates

central Saharan Tibesti Mountains, can be subdivided into four formations, each with a distinctive array of landforms:

- The Basal Sandstone, which is a very homogeneous, medium-to-fine grained rock, splits easily along bedding planes to give benched outcrops.
- The overlying Quatre-Roche Sandstone, which is characterized by inter-bedded conglomerate, cross-bedding and closely spaced joints, has been weathered and eroded into a maze of rock towers.
- This is not so in the Tabiriou Sandstone, which is characterized by frequent changes from peltic to arenaceous beds.
- The uppermost formation, the coarse-grained to conglomeratic Eli-Ye Sandstone, has been cut into a landscape of deeply incised gorges which generally follow the dominant fracture patterns.

It may be argued that diversity such as that reported by Roland can be dismissed simply as a matter of the scale of observation (that such variation is minor within the overall pattern characterizing a morphoclimatic zone) but that line of argument does not hold when we turn to other difficulties with the climatic hypothesis.

As we have already seen in an earlier chapter, and as Mainguet (1972) had previously pointed out, major clifflines are by no means limited to arid and semi-arid lands. Moreover, as Mainguet also emphasized, rounded or blunted forms

Fig. 8.4 Rounded slopes have formed above critical partings (bedding planes) in the De Chelley Sandstone near Monument Valley, Arizona, USA. Jointing maintains straight cliff faces; and the curved fracturing hackle marks show the failure of blocks off one face was initiated at the top of the cliff, and generated downward

(*forms émoussées*) are common on the sandstones of arid lands (Fig. 8.4). Indeed, rounded forms may occur in juxtaposition to angular cliffs. The convex summits of the great cliffs and towers of the Hombori and Tassili in the Sahara are a case in point (Beuf *et al.*, 1971; Mainguet, 1972, Plate LXIV); other instances can be cited from the Colorado Plateau – notably from the Navajo and Wingate Sandstones, the Aztec Sandstone of Nevada – and from central Australia – Uluru and Kata Tjuta. Although these rounded forms can be explained by structural and lithological constraints, some authors have interpreted them in terms of climatic fluctuations over geological time.

Bremer's interpretation of Uluru has been discussed earlier in Chapter 4, but probably the most thoroughly argued and best-known claim for climatic control of varied slope form in dry regions is in the account by Ahnert (1960) of the scarps of the Colorado Plateau. In fairness, it must be said at the outset that our criticism of the accounts by Bremer and by Ahnert relies on a far greater range of evidence for diverse morphology in dry lands than was available 50 years ago. Ahnert argued for differing forms reflecting differing climates:

- Vertical cliffs were developed by the dominance of basal sapping during pluvial periods.
- Rectilinear slopes were formed by the dominance of surface wash during drier phases.

- Rectilinear slopes truncated at the base by cliffs were the result of a period of wash giving way to one of sapping.
- Rounded slopes were the result of a phase of sapping being followed by one dominated by wash.

However, as we have shown in Chapters 3 and 4, the constraints of structure and lithology offer a far more plausible explanation for scarp form than a complicated scheme of climatic change. Furthermore, if Ahnert's assertions were sound, there ought to be a very widespread systematic distribution of the changes from one slope type to another, for climatic fluctuations of sufficient magnitude and duration to change bedrock forms must have occurred at a regional scale. But the detailed descriptions of scarps across large parts of the Colorado Plateau (Oberlander, 1977, 1989; Howard and Kochel, 1988) show that changes in morphology are localized rather than spatially systematic at the regional scale.

This is not to say that climatic change has had no effect on sandstone topography in what are now arid lands. Mainguet (1972) has shown how relic lateritic crusts on the summits of mesas in the Ennedi region of Chad mask fissures that are being picked out in great detail by erosion further downslope. Furthermore, Hagedorn (1971) demonstrates in parts of the Tibesti that fossil valleys and pediments that were cut in previously wetter periods are aligned obliquely to the orientation of modern dunes and of wind-cut furrows and yardangs in sandstone. The most impressive examples of bedrock relics of past climates are, however, the numerous sandstone karst features that occur in the Sahara. Complex fields of towers, no less impressive than those of the seasonally dry tropics of northern Australia, occur in Tibesti, Borkou, Tassili, Hoggar and Djao (Furst, 1966; Mainguet, 1966, 1972; Beuf *et al.*, 1971; Busche and Erbe, 1987; The Natural Arch and Bridge Society, 2008). Furst (1966) presented superb illustrations of arrays of towers and spires that rise abruptly from the flat serir of Tibesti. Busche and Erbe (1987) documented an impressive array of karst features in the sandstones of the Djao Plateau and Kaouar escarpment, including large, circular solution hollows, scarp-foot depressions, and extensive cave and tube systems. Many of these features are now being buried by aeolian sands, so they cannot have formed under the contemporary arid regime. Busche and Erbe argued, from evidence of past climates in the Sahara, that the varied forms of sandstone karst must be relics from seasonally humid climates of the Miocene or perhaps the Pliocene.

Osborn (1985) argued that the spectacular ruiniform landscape of southwestern Jordan was also carved during wetter climates earlier in the Cainozoic, and that it has undergone little change since the advent of aridity. This landscape consists of low-angled plains separating cliff-lined towers with rounded summits that rise up to 800 m above sand flats (Fig. 8.5). Although there are numerous scars caused by rock falls from the cliffs, there is little talus because the weakly cemented

Fig. 8.5 Rounded towers above sandy desert in Jordan. (Photo: P. Migon)

sandstone disintegrates on impact. More recent accounts of this general area have drawn attention to weathering under arid conditions. Goudie *et al.* (2002) described in detail the remarkable columnar weathering networks that they attribute to salt weathering in alcoves and case hardening on the intervening columns. They also drew attention to structural and lithological constraints, for cavernous weathering is best developed on surfaces facing opposite to the dip; and caverns in the white sandstones grow vertically, while those in the red sandstones grow horizontally. They pointed out that, as there is little cavernous weathering on the scars left by rock falls, the weathering must be proceeding very slowly or be very old. Measurements of Nabatean monuments in sandstones at Petra certainly demonstrate slow weathering (Paradise, 2003). The surfaces of the monuments are disintegrating at maximum rates ranging from about 20–80 mm/ 1000 years, but with marked differences according to aspect. The slowest retreat is on north-facing surfaces which receive less sunlight and are partly protected by lichens; and retreat is faster on east-facing and west-facing surfaces which are subject to the greatest changes in surface temperatures.

Wind abrasion in hyper-arid regions

Although aeolian sand blasting plays a role in the development of some minor landforms, the important role once attributed to aeolian erosion in arid and semi-arid lands now seems to have been exaggerated. As Gregory (1938) pointed out

70 years ago, caves in the Navajo Sandstone – once attributed to sand blasting – are used by 'man and beast' to shelter from sand storms, and are actually the product of chemical breakdown of the rock (see Chapter 6). Wind abrasion features are usually small-scale. For example, Youngson (2005) described ventifacts and pavements on silcrete in Central Otago, New Zealand, forming under intense westerly winds and in the rainshadow of the Southern Alps. Yet the dominance of sand blasting in shaping large areas of sandstone terrain in the hyper-arid parts of the Sahara cannot be denied. Mainguet (1966, 1970, 1972) drew attention to the truly astonishing furrowed landscapes in the sandstones of Borkou. Yardangs cut in sandstones are up to 20 m high, 200 m wide and 1 km long. Rock bars separated by grooves or passageways in this region display a rigorously parallel pattern aligned in the direction of the dominant winds. Individual bars extend for distances up to 22 km. Furthermore, as Hagedorn (1971) also observed in the Tibesti region, the wind-eroded furrows of the Borkou are aligned at about 90° to the remnants of fossil drainage lines. Mainguet (1970) concluded that aeolian corrosion not only obliterates landforms shaped by water action, but also creates a landscape that conforms to stationary wave laws. Wind erosion of sandstone, especially the cutting of extensive fields of yardangs, is also known from the Nubian Sandstones of Sinai and other parts of Egypt; and Vincent and Katton (2006) have recently described yardangs cut in the Saq Sandstones of northwest Saudi Arabia. Localized erosion of sandstone by wind may occur elsewhere, but the dominance of wind action on this scale seems limited to hyper-arid environments.

Sub-tropical lands

The landforms of the Carnarvon Range of sub-tropical central Queensland, formed in the Jurassic Precipice Sandstone, lie in a region which is now sub-humid (620–700 mm/year of rainfall), but would have been much wetter through most of the mid-Tertiary when the region supported rainforest. Basalts on the summits and valley floors testify to long development of the landscape, stretching back well before basalt extrusion 25–35 million years ago (Young and Wray, 2000). As described in Chapter 7, long cliff-lined escarpments, slickrock slopes, streams aligned down dipslopes and solutional features characterize the sandstone. Hence, the landscape has elements of several morphological elements often thought to be typical of other climatic regimes.

Probably the most impressive and well-researched sandstone terrain in the subtropics lies around Mt Danxia, on the southern side of the Nanling and Wuyi Mountains in Guandong, southeastern China (Peng, 2000, 2007; Zuo Dakang and Xing Yan, 1992). Average annual rainfall is 1450–1850 mm, with a strong summer maximum; and average temperatures range from 18 °C in summer to 5 °C in winter. Average elevation is 300–400 m, but peaks in Danxia reach above 600 m; and in similar terrain in the Taining region of Fujian, Dayading Peak reaches 913 m.

The Late Cretaceous continental red beds, into which this terrain is cut, consist mainly of well-cemented sandstone and conglomerate, with minor amounts of siltstone. While these rocks support a rugged cliff-lined landscape, the weaker underlying sandstone and siltstone give rise to hilly terrain. The typical slope complex around Mt Danxia consists of three main units:

- The slightly rounded summits are formed mainly by surface spalling of the sub-horizontal bedding and also by the angle of internal friction of the residual weathered debris.
- The prominent cliffs are controlled by major vertical jointing. The collapse of joint-bounded blocks is the main form of failure on the cliffs, though weathering along bedding planes is also significant.
- The footslopes, which are generally inclined at about 30°, seem to be controlled by the angle of internal friction of their debris mantle. However, massive collapse from the cliffs has in places scattered blocks as large as 30 m^3 across the footslopes and into the channels of streams.

The combination of cliffs, slot canyons, waterfalls and numerous sandstone towers gives a ruiniform appearance. Large-scale cavernous weathering is prominent on many cliff faces of this highly porous sandstone. So too are solution runnels, especially the incredible array of parallel vertical runnels that descend over the entire face of Shaibu Rock in the Wuyi Mountains.

The array of landforms around Mt Danxia is thought to have developed during the last 6 million years, and the major escarpments seem to be retreating at about 50–70 m/Ma. Peng (2000) outlined stages in the development of this landscape – regional faulting was followed by erosion down major joints that were then gradually widened to give the present array of landforms; and thereafter valleys continued to widen to form a planation surface with scattered residual inselbergs.

The reports prepared for the application for World Heritage status (Construction Department of Hunan Province, 2007; Peng, 2007) have shown the sandstone terrain of southeastern China to be complex, with considerable variation from the angular ruiniform landscape of the Danxiashan. The sandstones and conglomerates of the Langshan and Wanfoshan of Hunan Province have been eroded along closely spaced joints to form complex assemblages of towers (Fig. 8.6). But, whereas the towers of the Langshan have a distinctly rounded morphology, those of Wanfoshan have a pyramidal morphology. And in contrast to the close spacing of towers in both of these areas, those of the Jianglayshan in Zhejiang Province are widely spaced, and in some places quite isolated from other towers. Although the Chishui sandstone terrain of Guizhou Province also has some towers, it is

Fig. 8.6 The Danxia-style towers of Wanfoshan, Hunan Province, southern China

comprised mainly of extensive cliff-lined plateaus and mesas that bear a striking resemblance to other sandstone regions such as the Blue Mountains of temperate southeastern Australia. Variations in lithology, especially the relative importance of sandstone, siltstone or conglomerate, and also in structure and the intensity of jointing, seem to be the major constraints on these differences in morphology.

Temperate zones

Lands with Mediterranean climate

The most extensive sandstone landscapes in the temperate zone of winter-dominant rainfall occur in South Africa. Particularly rugged terrain has been cut from the thick sequences of arenites, conglomerates and siltstones of the Table Mountain Group. At Donkerkloof in the Cedarberg Range, for instance, an 850-m exposure of members of the Table Mountain Group displays multiple clifflines and benches (Rust, 1981). The thickly bedded and strongly cemented quartz arenites of the Peninsula Sandstone in the lower part of this exposure form massive cliffs that contrast to the multiple faces and small ledges of the upper cliffs cut in the thinly bedded sandstones of the Nardouw Formation. The two sets of cliffs are separated by prominent benches, and also debris slopes, on shales and tillites. Moon and Selby (1983) have quantified the relationship between rock mass strength and slope inclination in the 500-m high scarps in the quartzitic sandstones of Table Mountain, and in the quartzitic greywackes of the Clarens Formation on the footslopes of the Drakensberg (Fig. 8.7). They demonstrated that the generally high mass strength of sandstones in both areas is capable of supporting a complex of long-term equilibrium slopes that includes vertical cliffs and other steeply inclined segments. As noted previously in Chapter 6, the sandstones of South Africa also display a considerable range of solution features ranging from pans and runnels to large cave systems. The development of cliffs, platforms and solutional features at various scales in the Clarens Formation of the Drakensberg massif has been described by Twidale (1980). Twidale commented (p. 219) that the Clarens Formation was 'formerly and evocatively known as the Cave Sandstone'.

In southern Western Australia the ridge and valley terrain of the Stirling Ranges rises abruptly from the surrounding lowland. The ridges, which are cut in Precambrian quartzite, have multiple cliffs on their upper slopes. This upland is an ancient feature that has undergone very slow change, for it rises high above the nearby Yilgarn Plateau on which there is an extensive paleodrainage network of Eocene age. In a strikingly similar fashion, the Grampians rise abruptly from the plains of western Victoria. This rugged landscape was cut in generally gently

Fig. 8.7 Cliffs in the Clarens Formation sandstone, Lesotho, southern Africa. (Photo: C. R. Twidale)

folded Devonian and Carboniferous sandstones, and had taken essentially its present form prior to the extrusion of basalts of Pliocene age across the surrounding lowland. The main scarps are bounded by major clifflines; numerous slot-like canyons and corridors extend along the prominent regional joint sets; and tessellation, runnels and cavernous weathering features are common.

Various examples from Europe also show that the Mediterranean climatic zone is conducive to the development of steep, but variable slopes on sandstone, and that solutional features often occur. The steep-sided towers of Meteora in Greece are a striking example. In eastern Spain, erosion of prominent bedding and jointing in the gently dipping Albarracin Rodendo sandstones has created a tower and ruiniform relief, in which there are numerous solutional features (Lozano *et al.*, 2008). And the erosion of intensely jointed conglomerate and sandstone has formed the spectacular rounded towers at Montserrat.

Humid warm temperate lands

Tricart (1979) warned of the bias inherent in generalizations about the geomorphology of temperate humid lands drawn from research in the Northern Hemisphere where climatic oscillations, especially during the Quaternary, were much greater than those that occurred over most of the Southern Hemisphere. He suggested that studies in the Southern Hemisphere might resolve ambiguities in our knowledge of the effects of climatic change on the long-term evolution of landforms. These ambiguities may be of very considerable proportion if – as Büdel (1982) suggested – more than 95% of the relief features of central Europe are not the outcome of the landform processes governed by the present climate but are relicts of previous climates.

The sandstone lands at the southern end of the Sydney Basin, in southeastern Australia, provide an exceptional opportunity for determining the development of morphology over extremely long periods of relatively stable climate. This region has had a humid climate for some 50 million years. Macrofossils and pollen in sediments under basalts of Middle Eocene age (Wellman and McDougall, 1974; Young and McDougall, 1985) demonstrate that the region was then covered by warm temperate rainforest (Martin, 1978). The pollen record for southeastern Australia shows that humid conditions continued until at least the Middle Miocene (Martin, 1978). This fact has been confirmed for the Shoalhaven region by the presence of macrofossils and pollens indicative of cool temperate rainforest (Nott and Owen, 1992) in lake sediments that can be traced under Oligocene basalts, and also by the presence of cool temperate rainforest fossils in sediments proven to be of Miocene age in adjacent catchments (Martin, 1978). The drying of the continent during the Late Miocene and Pliocene led to most of the rainforest in this region being replaced by eucalypt woodland and forest, but rainforest species still survive in sheltered valleys, for average annual rainfall over most of this sandstone country is more than 1000 mm, with some areas receiving about 1800 mm. The average annual runoff over most of the sandstone country is estimated to be more than 300 mm, with substantial parts yielding over 800 mm. Moreover, the effectiveness of the climatic regime is not offset by water loss into a permeable rock. Water yield is mainly from overland, rather than from deep subsurface flow, and the low permeability of these sandstones is demonstrated by their ability to support extensive, permanent sedgelands on most flat summits. It should also be noted that this region was not subject to periglacial action during the Pleistocene (A. R. M. Young, 1986).

This is an extremely rugged landscape of mesas, cliffs and canyons. The uplands are being dissected by the headward retreat of canyons, many of which terminate in major waterfalls; and stream profiles in the upper sections of the canyons are broken by numerous rapids and small falls. The virtually continuous clifflines are shaped mainly by the collapse or gliding of joint-bounded blocks. Where fracturing is pronounced, ruiniform relief is well developed. Blunting or rounding of the summits of outcrops in the southwest of the region is clearly related to pebbly or closely bedded facies; these rounded forms occur adjacent to

Fig. 8.8 The plateau of the upper Clyde Valley, southern Sydney Basin. The cliffs of the mesas are in Nowra Sandstone; Monolith Valley is the upland gap between the mesas. The lower Snapper Point Formation cliffs lie unconformably on Siluro-Devonian Lachlan Fold Belt beds

the bold near-vertical cliff faces of the more massive facies beneath them, and in many places are truncated by fractures extending upwards from failures along joints (Fig. 8.8). The most obvious effects of chemical weathering are the numerous caverns, or tafoni, on the faces of cliffs (see Chapter 6), but large-scale karst features are rare. Upland valleys are shallow, largely filled with sediment, generally lack continuous channels over much of their length, and support dense covers of sedges and shrubs. As described in Chapter 7, their form is strikingly similar to that of the dambos of the tropics, but the Late Pleistocene and Holocene ages of the sedimentary fills show that these valleys are in equilibrium with contemporary climates, and are not relics of previously warmer regimes.

Despite the longevity of humid temperate conditions and the general spatial uniformity of climate over the region, there is considerable diversity of morphology on the sandstones of the Sydney Basin, mainly because of structural and lithological variability:

- On the more massive sandstones in the south, the clifflines are rarely notched by tributary streams, and the interfluves are broad and undulating.
- In the north, where the sandstones are generally more thinly bedded and intercalated with numerous clayey or silty beds, the terrain is extensively dissected.

Fig. 8.9 Broad pediments lie between the cliffs of the dissected plateau and the broad benches on lower sandstones in the Capertee valley, west of Sydney, Australia

- In the Blue Mountains on the western side of the Basin, where major claystone beds are intercalated with the sandstones, the cliffs are notched by hanging valleys that produced considerable relief on the interfluves.
- Most canyon floors throughout the Sydney Basin are narrow, but on the western side of the Blue Mountains, where streams have cut down into thick shales, the floors of canyons are up to several kilometres wide. Valley floors in the upper Capertee River catchment take the form of broad pediments sweeping down from isolated sandstone mesas, producing a landscape reminiscent of the Colorado Plateau (Fig. 8.9).

These contrasts over a distance of only about 100 km provide further illustration of the difficulties inherent in the use of single type examples for morphogenetic comparison. The differences in form of interfluves and slopes over this climatically stable temperate region mimic those contrasts widely ascribed to the effects of humid versus arid regimes.

Humid cool temperate lands

From the preceding discussion, it is now clear that Hettner was correct in rejecting the uncritical assumption that the sandstone mesas of Saxony must be

Fig. 8.10 The mesa-dominated landscape of the Elbe region, Saxony. (Photo: P. Migon)

relics of drier climates. Not only do they resemble the landforms of arid and semi-arid lands, or even those of some parts of the seasonally dry tropics, but they are strikingly similar to landforms in the southern and western Sydney Basin which were formed under humid temperate climates. Flat-topped mesas such as the Lilienstein, or the vertical, jointed walls above the Elbe near Bad Schandau (Fig. 8.10), would not be out of place either in the Capertee or Shoalhaven catchments; the great jointed towers and pinnacles of the Bastei Plateau, especially those near Basteibrucke, are virtually identical to those in and around Monolith Valley on the divide between the Shoalhaven and Clyde Rivers. And, like the outcrops at Monolith Valley, the Saxon sandstones are surprisingly weakly cemented (Robinson and Williams, 1994); the height of the cliffs seems more related to the thickness of individual beds than to rock strength. In short, it could be argued that the sandstone landscape of Saxony is essentially characteristic of the morphogenetic regime under which it is now found. This landscape certainly cannot be written off as an unusual occurrence in otherwise dominantly rounded or subdued sandstone forms in central and western Europe. Sandstone landforms further east in the Bohemian Basin, especially near Adrspach-Teplice 'rock city', are also characterized by bold, rather than subdued outcrops; and they include many prominent, vertical cliffs and tall, jointed-bounded spires (Jerzykiewicz and Wojewoda, 1986; Varilová, 2002) (Fig. 8.11). So far from lithological or structural variations being subdued by climatic effects, these outcrops display a very detailed control of topography by depositional features such as giant cross-beds, and by preferential

Fig. 8.11 A 'rock city' – the Bohemian Paradise – with tall spires separated by broad corridors or 'streets'. (Photo: P. Migon)

weathering of the ancient erosional planes transecting sets of cross-beds. Solutional features are also found, as they are in the Fontainebleau Sandstone in France. Again, the form of these smaller-scale features in Europe is akin to the form of the features in humid southeastern Australia. Arguing on the basis of form alone that they must have formed under an arid climate is not tenable.

Control of morphology by jointing and bedding is also prominent in many English sandstones, notably the Millstone Grit of the Pennines and the Ardingly Sandstone of southeastern England (Robinson and Williams, 1994). In central USA, an array of similar forms occurs in the now cool temperate Driftless Area of Wisconsin. In the Driftless Area, which was subject to periglacial but not glacial action, sandstone outcrops rise abruptly from the surrounding lowlands, forming vertical walls and turrets like Castle Rock, north of Madison (Thornbury, 1965).

The role of Pleistocene periglaciation in Europe

Given the similarities described above between landforms in different modern climatic regimes, what was the impact of Pleistocene periglaciation on the sandstone terrain of these cool temperate lands? Mainguet (1972) argued that, in places, periglacial effects were considerable; and that the sandstone terrain of the Vosges in northeastern France – which both she and Tricart and Cailleux (1972) used in their comparative studies – is not representative of temperate climates. She pointed especially to periglacial effects in the forming of the broad, alluviated valleys, the generally regular stream profiles, and the numerous but small structural steps on the hillsides. Tricart and Cailleux made similar observations concerning the landforms of the Vosges, noting in particular that, although abrupt escarpments are rare, hillsides with slopes of 20–30° are not exceptional. Blume and Remmele (1989) described similar modifications of the Bunter Sandstone scarps in the Black Forest by the cutting of minor cirques and the accumulation of thick periglacial deposits of boulders. However, if periglaciation was the dominant cause of the subdued scarps on the sandstones of the Vosges and Black Forest, why are the cliffs of the Saxon sandstones so prominent when they must have been subjected to the same process? This contrast in the prominence of cliffs, emphasized in Hettner's critique, seems to reflect fundamental differences in lithology and structure. Although the 500-m thick Triassic sandstones of the Vosges contain some very resistant beds, the irregularity of the stratigraphy caused by numerous faults, together with the presence of poorly cemented sands, precludes the development of a bold cuesta topography (Tricart and Cailleux, 1972). Furthermore, the outcrops of the cliff-forming beds in the Black Forest are often quite thin (Blume and Remmele, 1989). These initially more subdued outcrops probably were more susceptible to masking by periglacial processes than were the cliffed mesas in Saxony, but were not primarily sculpted by periglaciation.

In some places, the impact of past periglaciation is clear. Coxon (1988) described periglacial forms exhumed from beneath peat beds on the Truskmore plateau in Ireland. The plateau is capped by the horizontally bedded Glenade Sandstone, which is a massive, current-bedded, medium-to-coarse sandstone. Networks and sorted circles of blocks of the Glenade Sandstone, set in a silty sand matrix, occur on the summit of the plateau. Frost-shattered blocks of this sandstone extend down the flank of the plateau, forming extensive stone stripes with impressive frontal lobes. The size of the features seems indicative of severe cold with frequent freeze–thaw cycles, differing considerably from the present-day cool maritime climate.

Assessing the degree of widespread previous periglacial influence, however, is problematic in the sandstone landscapes of England and Wales. Even though large areas of uplands now have undulating surfaces blanketed by peat, most of the main sandstones – such as the Millstone Grit of the Pennines and the Barmouth Grit of the Harlech Dome – are sufficiently massive, strongly cemented and jointed to support vertical faces, and we need not invoke periglacial action to explain them (Sparks, 1971). Nevertheless, angular tor-like blocks and pinnacles, and deposits of very coarse rubble (Sparks, 1971, Plate 40), have been attributed to periglaciation (Palmer and Radley, 1961). These tors are not unweathered remnants or corestones exhumed by the stripping of deep weathering profiles, but are chiselled out of hard rock. They invariably rise out of rock rubble and soli-fluction material along escarpments; they never stand alone on the extensive summits. The Stiperstones of Shropshire, for example, are cut from quartzite outcrop standing above a debris slope on which the stripes and polygonal structures in angular debris are indicative of former periglacial action. Palmer and Radley argued that tors like these were formed by diurnal freezing and thawing in joints exposed in the active layer above permafrost, and may therefore be regarded as 'Paleo-arctic' rather than 'Paleo-tropical'. A periglacial origin for them is supported by the present-day shaping of similar features in the Beacon Sandstone exposed in Wright Valley of Antarctica (Derbyshire, 1972). The Wright Valley tors, however, are being formed by frost shattering above ground in a polar desert regime, not below ground level in the water-abundant active zone above permafrost. From their comparative study of four sandstone land-scapes in England, Robinson and Williams (2005) found no clear relationship between weathering features and their glacial history. Bare pavements are uncommon, and pillars and tors occur within areas covered by Devensian ice. Block fields and clitter slopes probably formed under intense periglacial condi-tions, but again the angularity of cliffs seems more related to rock hardness than to climatic history.

The role of glaciation

The effects of Holocene periglacial and fluvial action notwithstanding, the imprint of previous glacial erosion of lithologically and structurally variable sandstones is preserved clearly in the Scottish highlands (Godard, 1965). The arkosic Torridonian Sandstones of northwestern Scotland are thickly bedded, well jointed and strongly cemented. During the Quaternary, the dissected land surface on these sandstones was over-ridden and extensively eroded by glaciers (Fig. 8.12). Most of the upland remnants were carved into precipitous pyramidal peaks and serrated aretes; Suilven, Canisp and Stac Polly, north of Ullapool, are outstanding examples. Numerous deep cirques were excavated, especially in the vicinity of Beinn Eighe and Loch Torridon. Valleys were deepened and their sides steepened, and now form spectacular fjord and finger lake systems such as Loch Broom and Loch Maree. In short, the surface of the Torridonian Sandstone was transformed into a classic ice-scoured landscape which once must have closely resembled the modern Ellsworth Mountains of Antarctica.

Fig. 8.12 Steep cliffs and scree slopes on the highly cohesive sandstones and quartzites, and coarse angular moraine below them, show clear evidence of past glacial and periglacial erosion in the Torridonian Mts, northwest Scotland

Despite the extensive glacial erosion, the structural and lithological imprint of the sandstone is still quite clear. Although precipitous with much rocky outcrop, most slopes are broken by stacked benches; these are lithologically controlled and follow variations in the dip of the prominent bedding. Benches are particularly prominent on the walls of Loch Broom opposite Ullapool, where they dip up-valley, and on the north flank of Glen Torridon, where they are almost horizontal. Gordon (1981) has demonstrated in the Torridonian mountains around the Kyle of Lochalsh that the major features of the landscape are still structurally controlled. The landforms of that area are mainly aligned northeast–southwest along the strike, or east–southeast along the dip of the sandstone, rather than in the direction of the ice which moved to the north and northeast. What is more, Haynes (1968) has shown that the morphology of cirques cut in the Torridonian varies with geological and topographic factors, as well as with duration and type of glaciation. Jointing and bedding strongly constrain the depth of the cirques and the form of the flanking cliffs. Although the sandstones are massive, thin shale partings along many bedding planes were readily stripped by the ice so that the erosional base of the cirque glaciers paralleled the dip. Where the dip is towards the head of the corrie, rock basins, now occupied by small lakes, were excavated. Where the dip is towards the mouth of the cirque, very few reversed slopes developed, and hence lakes did not form (Haynes, 1968).

The glaciated landscape of northeastern Scotland, especially on the Old Red Sandstone of Caithness and the Orkney Islands, seems the very antithesis of the rugged Torridonian landscape. The Old Red Sandstone has been carved into a spectacular coastline of huge cliffs and stacks, deep caves, arches and chasms (*geos* – see Fig. 7.4), but the hinterland consists mainly of a rolling, low plateau (see Fig. 3.8). Only on the Orkney Islands, especially on Hoy, is there substantial local relief; and there the land surface is dominantly smooth, with remarkably few rocky outcrops. Away from the coastline, the terrain is far more like that normally associated with weak argillaceous rocks than with sandstones, and the immediately obvious evidence of glaciation is limited to the rock basins of a deranged drainage pattern and to prominent moraines near Rackwick on Hoy.

The great contrasts in the glaciated sandstone terrain of northern Scotland are due to a combination of glaciological and lithological factors. Although the ice cut deeply into the high western highlands, it had a much smaller effect in the northeast. Tors on the Cairngorm Mountains (Sugden, 1968) and preglacial weathering profiles and sediments in Buchan (Hall, 1986) survived the glaciation of northeast Scotland, thereby demonstrating that, away from deep valleys, the erosional effect of the ice cap was slight.

Lithological characteristics also contribute to the subdued nature of the Caithness and Orkney landscapes. The Old Red Sandstone thereabouts is very thinly bedded, and in the Caithness Flagstone Series, on the northern Scottish coast, sandstones alternate with mudstones, bituminous flagstones and calcareous beds (Sparks, 1971). These very thin flaggy beds disintegrate rapidly, producing a pronounced smoothing of slopes only a few tens of metres behind the intricately stepped outcrops exposed at the top of the sea cliffs. The Flagstone Series is not present in the Orkney Islands, but the relatively thin bedding which persists in the Old Red Sandstone over most of the islands gives rise to smooth slopes. Benched rocky outcrops occur on some hillsides in parts of Hoy, Mainland and Rousay where there are apparently more massive beds. Unlike the high and dissected lands on the Torridonian Sandstones in the west, the smooth, undulating terrain, and generally low elevation of this region was not conducive to deep glacial erosion by ice.

The relative importance of glaciation and pre-glacial fluvial erosion apparently varies from place to place. Glaciation played the dominant role in shaping the Sogne Fjord of western Norway, though much of the pre-glacial land surface survives on the adjacent plateau (Nesje and Whillans, 1994); and the same relationship presumably holds in the nearby sandstone terrain. On the other hand, most of the denudation of greywackes near Mt Cook, in the Southern Alps of New Zealand, was accomplished by pre-glacial fluvial action, with the estimated lag time for conversion to glacial troughs being 200–300 Ka (Kirkbride and Matthews, 1997).

Polar lands

Although all polar lands are subjected to extremely low temperatures, at least in winter, there is considerable climatic variation within this zone. Polar maritime areas generally have more moderate climates, with a smaller seasonal range of temperature, than polar continental areas. They also tend to receive higher precipitation. Differences in sea temperatures may also produce considerable variations between maritime areas. For example, Spitzbergen, at the northern extremity of the Gulf Stream, has average daily temperatures of about 4 °C in summer and −13 °C in winter; while the comparable temperatures at McMurdo in Antarctica, which lies at a similar latitude, are about 0 °C and −23 °C. And whereas Spitzbergen receives an average annual precipitation of about 340 mm, McMurdo receives only 200 mm. Moreover, there is a sharp climatic gradient to colder and hyper-arid conditions inland from McMurdo across the Dry Valleys region. Sandstones in polar lands can thus be subjected to different kinds of glacial or periglacial erosion, to post-glacial stress release, salt weathering and biochemical weathering. Differences in the properties of sandstones, especially porosity, fracturing, stiffness and Rock Mass Strength also have significant effects on the development of landforms under low temperature regimes.

Much of Spitzbergen consists of an active alpine-type glacial landscape, with dissected summits reaching elevations of around 1400–1700 m. It lives up to its name, bestowed by Willem Barents, which means 'jagged peaks'. Although ice still reaches tide water, the extensive glacial troughs carved during glacial maxima are now occupied largely by fjords. Woodfjord, in the north-central part of the main island, is flanked by aretes and peaks cut in the Devonian Old Red Sandstone (Fig. 8.13); Kongsfjord on the west coast is partly flanked by sandstone and conglomerate; and the southern side of Isfjord, near Longyearbyen, is cut into sandstone, shale and coal beds. The glaciers now are mainly polythermal, being cold-based in winter but wet-based with meltwater in summer. Rock faces from which the ice has retreated appeared to be modified by three main processes of very variable rates of change, that Andre (1997) referred to as a triple-rate evolution. Flaking caused by lichens proceeds very slowly at about 2 mm/Ka; frost shattering proceeds at a moderate rate of about 100 mm/Ka; and mechanical splitting caused by post-glacial stress release is by far the fastest, proceeding at about 1 m/Ka. However, vertical walls of massive quartzite along Kongsfjord have retreated no more than 1–2 m during the Holocene, and this retreat has been due mainly to stress relaxation (Andre, 1997).

The German research expeditions to Barentsoya in southeastern Svalbard were aimed at not only increasing knowledge of cryoturbation processes, but also at investigating the processes that result in exceptionally wide valley floors and 'the

Fig. 8.13 Glaciation still affects the peaks in Old Red Sandstone along Woodfjord, Spitzbergen

extraordinary down-cutting capacity of non-glacial cold stage rivers' (Büdel, 1982, p. 49). The plateau surface of Barentsoya is cut in almost horizontally bedded arkosic sandstones and dolerites, and, with maximum elevations little more than 400 m, is much lower than the peaks on Spitzbergen. During glacial maxima, this plateau seems to have been covered only by thin, barely moving, ice caps that had little erosive effect, and may even have protected the land surface from frost action (Büdel, 1980). Büdel gave a detailed account of the very well-developed cryoturbation features, such as stone polygons and circles, and soli-fluction sheets on the arkose exposed on the plateau surface (Fig. 8.14). Büdel's major contribution was to draw attention to the role of the layer of seasonally frozen debris, which he termed the 'ice rind', at the top of the fragmented arkose on Barentsoya. It is the disruptive effect of the ice rind, he argued, that is the driving force of the 'excessive valley-cutting' typical of many periglacial areas. Mercier (2002) has proposed the term 'paraglacial', rather than periglacial, for the erosion of rills and gullies on steep slopes in sandstone and shale near Longyearbyen, but of course such erosion by meltwater is a seasonally significant

Fig. 8.14 Frost-shattered Old Red Sandstone and stone polygons in Spitzbergen

agent in all periglacial lands. Büdel certainly had already noted the importance of running water on Barentsoya, firstly by filter drainage of finer material carried by meltwater through the gaps between coarser clasts, and then by the transition from solifluction to rill wash and gully erosion where slopes exceed 25–30°. Even more striking examples of gullying on steep slopes under periglacial conditions occur in the sequence of Cretaceous and Tertiary sandstone and mudstone on the Nuussuaq Peninsular in western Greenland (see Fig. 2.14).

In striking contrast to the very active denudation observed in Svalbard, the ice-free areas on the edge of Antarctica appear to be essentially ancient relict landscapes that are undergoing minimal change under extremely cold and hyper-arid conditions. Cosmogenic dating indicates that the Amery Oasis in the northern Prince Charles Mountains records at least 10 Ma of landscape evolution (Hambrey *et al.*, 2007). Outcrops of sandstones and conglomerates of the Amery Group were eroded by wet-based, probably polythermal glaciers during Miocene and Pliocene times, but have been altered little by the mainly cold-based glaciers of Quaternary times. However, largely post-Pliocene uplift has exposed summits to intense periglacial action, though relicts of deep weathering and tor formation are preserved on some higher areas. Cosmogenic dating, together with the preservation of mid-Miocene volcanic deposits, indicates a similar history for the McMurdo Dry Valleys, where there has been minimal change to the landscape over at least the past 15 Ma (Summerfield *et al.*, 1999). Dating of surfaces on sandstone of the Beacon Supergroup, and also on the granitic basement, showed that sites with low local relief at high elevation are eroding at maximum rates of only 0.13–0.16 m/Ma, and rectilinear slopes at lower elevation are eroding at 0.26–1.02 m/Ma. Here too, the change of the East Antarctic Ice Sheet from a dynamic temperate state to a cold sluggish state seems to have been of prime importance.

On the western side of the Dry Valley region, there are huge relict drainage channels that C. A. Cotton (1966) termed the 'Antarctic scablands' because of their similarity to the channelled scablands of northwestern USA (Bretz, 1923). Instead of being formed by catastrophic flooding from a pro-glacial lake, however, the Antarctic channels were cut by floods from sub-glacial lakes (Sugden and Denton, 2004; Lewis *et al.*, 2006). The largest network of these scablands is cut through the sandstones of the Convoy Range, where channels and potholes cover a tract at least 50 km in length. Channels in this system are up to 600 m wide, and potholes are up to 140 m wide and 40 m deep (Sugden and Denton, 2004). Dating of a volcanic ash shower in a similar system cut through dolerite, 40 km further south, indicates that the cutting of the channels occurred some time between about 12–14 Ma (Lewis *et al.*, 2006). Since then there seems to have been minimal change in the landscape, apart from considerable post-glacial rebound (Stern *et al.*, 2005).

In the Ellsworth Mountains of Antarctica, ridges and nunataks carved from dominantly sandy sediments, that have been slightly to moderately metamorphosed, rise above the vast surrounding ice sheets (Rutford, 1972). On the eastern side of the Sentinel Range, one of the more prominent parts of the Ellsworth Mountain complex, numerous large cirques have bitten deeply into the flanks of the uplands, resulting in an arcuate ridge pattern. The other side of the same range has been carved into straight, parallel aretes separated by elongated valleys, rather than eroded by cirques. Rutford argued that these valleys are the result of modification of a pre-glacial, structurally controlled drainage pattern. He also asserted that most of the erosion by ice took place during an initial phase of valley glaciation, and that the later development of an ice cap had an essentially protective, rather than an erosive effect on the landscape. The striated and polished surfaces of surprisingly rounded summits on quartzites, in the central Heritage Range section of the region, leave no doubt of the general exhumation of the higher parts of the landscape from the ice sheet. Unless the ice sheet was essentially stagnant, as Rutford argued, it is difficult to see how the cirques, aretes and sharp peaks could have formed and survived.

Augustinus and Selby (1990) considered that the Olympus and Asgard Ranges of the McMurdo area were also shaped by valley glaciers, rather than by the Antarctic ice sheet (for a contrary view, see Denton *et al.*, 1984). They demonstrated that this largely sandstone terrain consists of two main types of slope assemblages. The upper slopes of the ranges have steep slopes in keeping with the mass strength properties of the sandstones and ortho-quartzites from which they have been carved. In striking contrast, the lower slopes have much gentler slopes than would be expected from the properties of the sandstones, and are interpreted as being controlled by the angle of repose of the thin cover of scree which mantles them (i.e. they seem to be 'Richter slopes'). Augustinus and Selby could not reconcile this very consistent pattern of strength–equilibrium slopes surmounting Richter slopes with deep erosion by an ice sheet.

The study by Augustinus and Selby of the Asgard Range clearly demonstrates the importance of variations in mechanical properties of rocks – specifically between orthoquartzites and less strongly bonded sandstones – in determining glaciated forms. The Beacon Orthoquartzite, which has a Rock Mass Strength rating of 83 units – mainly because of a high compressive strength and very wide spacing (100–200 m) of vertical joints – supports cliffs with an average declivity of 78°. The Altar Mountain Sandstone, which has a Rock Mass Strength rating of 69, supports equilibrium slopes of only about 58°. Augustinus (1992) has used a similar approach in accounting for the different morphologies of relict glacial troughs in New Zealand. Whereas the troughs in the crystalline rocks and schists of Fjordland on the west coast are deep and narrow, those in the greywackes near Mt Cook are wide with flat floors. Greywacke is the hardest of these rocks, with a point load strength twice that of the others, and has by far the greater resistance to abrasion. However, close joint spacing imparts a low Rock Mass Strength that facilitated glacial plucking.

It is significant that a failure on a 300-m high sandstone cliff on Mt Boreas was attributed to stress release, because it propagated upwards. In this extremely arid environment, water was not the trigger. At nearby Mt Circe, the same sandstone suffers horizontal stresses as a result of the gravitational load exerted by its own weight (Selby and Hodder, 1993)

A notable feature of weathering in the Dry Valleys of Antarctica is widespread occurrence of cavernous weathering in a variety of rocks, including sandstone. This has been attributed to processes ranging from salt weathering to capillary moisture flow (Conca and Astor, 1987), and its development is certainly facilitated by the hyper-arid climate. By way of contrast, Bromley and Mikulas (2002) noted that cavernous weathering is rarely seen in eastern Greenland, a fact which they attributed to insufficient evaporation of moisture from sandstone surfaces in a cold, though moister climate.

Conclusions

It is clear from the examples considered here that climate does exert an important influence on the shaping of sandstone terrain. But it is equally clear that considerable caution is needed in assessing that influence, and especially in attempting to formulate morphogenetic generalizations. Aeolian erosion of sandstone is important under hyper-arid conditions. Glacial and periglacial action can greatly modify sandstone surfaces. Very large-scale development of karst in sandstone is probably limited to the humid and seasonally humid tropics, although sandstone karst is found even in humid temperate lands. The question of duration versus intensity of weathering in the forming of these features remains problematic. Fluvially eroded sandstones are certainly distinct from those eroded by aeolian or glacial processes. Nonetheless, it is difficult to see any clear-cut means of subdividing this very broad, almost self-evident zonation. The climatic distinctions seem to be more subtle than, and over-ridden by, lithological and structural variability.

Even within these broad groupings, it is not possible – except perhaps for the hyper-arid zone – to identify type examples that are truly representative. The diversity within the groups is just too great. Which is representative of glaciated sandstones, Spitzbergen or Antarctica? Or if we decide to use both, is the subdivision really one of present or past climatic conditions? If we attempt a similar solution to the diversity of the tropics, using, say, the Cockburn versus the Bungle Bungle Range as type examples, the subdivision is unquestionably linked to variations of lithology rather than climate. Nor can this problem be circumvented with appeals to scale of observation, with climate controlling the broad pattern, and structure and lithology relegated to a secondary role. Roraima (in the humid tropical Venezuela) has more in common with Arnhemland in a savanna climatic zone, or for that matter with the arid area of Kiffa or the temperate Sydney Basin, than it does with Fouta Djallon in tropical Guinea. Consequently, even for a relatively homogeneous rock type like sandstone, we must dispense with assertions like that made by Büdel (1982, p. 2), who insisted that 'a natural system of relief formation must be based exclusively' on differences in relief forms governed by climate. The influence of climate cannot be denied, but the phenomena are far too complex for such a simplistic and rigid ordering. As Hettner foresaw, the classification of landforms by climatic zonation is inherently susceptible to circular argument – because the influence to be investigated becomes, from the outset, the basis for structuring the observations.

Reliance on type examples to characterize climatic zones without first investigating in detail the diversity within those zones led to premature generalization.

Steep faces are just as prominent in humid lands as they are in arid lands, and rounded slopes are common features of both zones, so slope form on sandstone is not a reliable guide to climatic history. Premature generalization led to the neglect not only of the interaction of erosive processes with the properties of the materials on which they operate, but also of absolutely fundamental instances of climatic influence on morphogenesis. Although karst in sandstone is widespread in the tropics, it figures prominently in none of the handbooks on climatic geomorphology, and at best is dismissed as 'pseudo-karst' of minor significance. One well-known handbook devoted to the tropics even asserted that 'quartzites are practically immune to chemical weathering' (Tricart, 1972, p. 152). Yet explaining how forms normally associated with rocks considered to be highly soluble, such as limestones, are also found in supposedly far less soluble rocks is obviously a major task in assessing the role of climate in the development of landforms.

It might be more profitable for us to follow the lead given by Twidale (1982) in his analysis of granitic terrain, and use the topography itself as the basis for classification. Mainguet (1972) went some distance in this direction in applying Hurrault's concept of *paysage regionaux* (regional landscapes) to sandstones, but topographic variation in diverse structural settings, as well as climatic settings, must be assessed before such an ordering of landforms can be achieved. An assessment must also be made of the place of particular landforms in a developmental sequence over time; in other words, the much-maligned concept of 'stage' can no longer be simply written off. This step is essential if only because the anti-Davisian diatribes that began many textbooks on climatic geomorphology were not adequate demonstrations of the frequently asserted tropical origin of erosional surfaces. Here, we turn now to the role of tectonic setting in the development of distinctive sandstone landscapes.

9

Tectonic constraints on landforms

An essentially geographical analysis of landforms, in which the explanation of variation from place to place is the central issue, must be soundly grounded on stratigraphy and tectonics. This much was made clear by our introductory example of the Bungle Bungle Range. In this, we demonstrated major changes in morphology with changes in lithology, and also illustrated the great morphological contrasts that can be found between adjacent, but markedly different tectonic settings. Here again, our prime concern is that 'the spacing of phenomena over the earth expresses the general geographic problem of distribution, which leads us to ask about the meaning of the presence or absence, massing or thinning of things or groups of things variable as to areal extension' (Sauer, 1941, pp. 357–358). The phenomena are geological, but the purpose is, in this sense, geographical. That is to say we are concerned with variability of landforms, not with stratigraphy and tectonics *per se*. We begin with the deformation of sediments during their deposition, and then turn to the varied styles of post-depositional deformation.

Syn-depositional deformation

Ever since W. Penck's critique, models of cyclical landform evolution have been modified to allow for a possible complexity of tectonic events during the course of denudation, but they still deal with tectonics as essentially post-depositional phenomena. Yet there is clear stratigraphic evidence not only for small-scale deformation (at the level of facies change), but also much larger-scale deformation as a basin fills with sediment. This is not just a matter of interest to sedimentologists and stratigraphers, but also to geomorphologists, because failure to distinguish syn-depositional from post-depositional earth movements can lead to serious errors in the understanding of the evolution of landscapes. The history of interpretation of landforms in the southern Sydney Basin provides an excellent example.

244

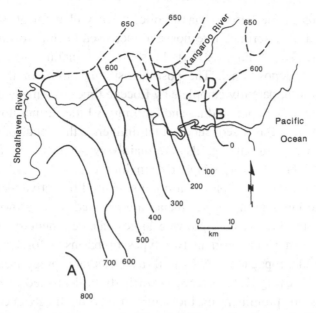

Fig. 9.1 Structure contours on the Permian Nowra Sandstone (*solid lines*) and Triassic Hawkesbury Sandstone (*dotted lines*) in the Shoalhaven valley, south of Sydney. The dip of the Nowra Sandstone from A to B was wrongly attributed to post-depositional warping. A syn-depositional origin for the structure is shown by the divergence of the Nowra and Hawkesbury Sandstones along the section C–D

Lester King's well-known account of the coastal flank of the uplands south of Sydney makes much of monoclinal, or cymatogenic warping during the Late Tertiary (King, 1959, 1962). His interpretation of the form, extent and age of the supposed warped pediplain surfaces, and of the tectonic events that displaced them, hinge on this concept. In fairness to King, who made a broad subcontinental reconnaissance as part of an intercontinental comparison, it must be said that he drew on earlier descriptions of this supposedly Late Tertiary monocline. Stratigraphic evidence, which was available even before the hypothesis for this post-depositional warping was first advanced, demonstrates that it simply did not occur. The inclined surface seen by King and earlier workers is, in fact, a stripped surface of Paleozoic age that can be traced northwards until it passes under a great mass of Late Permian and Triassic sediments. The increasing thickness of this mass of sediments exactly matches the fall in elevation of the basal Nowra Sandstone surface (Fig. 9.1). Near the former Permian shoreline, the Nowra Sandstone is near the top of the preserved sequence; farther east, near the modern shoreline, it is buried under more than 600 m of younger sediment. This landscape should not be interpreted in terms of successive pediplanations, greatly warped at the close of the Tertiary. Rather, it needs to be understood in terms of the deep erosion of the Permo-Triassic pile, the key element of which was the exhumation of a warp formed in response to increased loading towards the centre of the basin (R. W. Young, 1977a). In short, the inclination of the Nowra Sandstone is syn-depositional. The post-depositional uplift of the Shoalhaven region, which was apparently complete by Eocene times (Young and McDougall, 1985), produced only minor folding and faulting. A little further north, folds and faults in the Sydney Basin sequence, which influence the detail of drainage and slope patterns near Wollongong, also are mainly syn-depositional features.

In addition to the relatively simple deformation caused by loading of a basin, there is also much more complex deformation caused by active tectonics during deposition. The latter group of sediments are referred to as growth strata, and show a close interplay between growing large-scale sedimentary structures and deposition. Growth strata form in two types of basins – foreland basins that develop ahead of propagating fold-and-thrust belts, and piggy-back basins that form on top of moving thrust sheets. Growth strata deposited in both of these types of basins are prominent on the southern side of the Pyrenees and in the Catalan Ranges of northeast Spain (Verges *et al.*, 2002). Sandstones form prominent synclinal structures in the foreland setting of the Catalan Ranges, and the record of the interplay between longitudinal rivers and transverse alluvial fans on top of a piggy-back syncline is evident in the southeastern Pyrenees. Uplift on particular structures was generally similar to the rates of sediment accumulation (Verges *et al.*, 2002). Rather than deformation following deposition, as is commonly described in geomorphological textbooks, deformation during the compressive stage was here generally followed by tectonic quiescence during which the growth strata were buried. The present stage of stream incision has exhumed the older structures.

Post-depositional deformation at the micro- and mesoscale

Tectonically induced deformation often occurs in sandstone at a scale that is readily overlooked in the study of landforms. For example, the flexing of the Wingate Sandstone of the eastern Colorado Plateau has produced little major fracturing or faulting, but it has resulted in the generation of zones of dense micro-faulting. Micro-faulting in the Wingate Sandstone is conspicuous because of light-coloured, relatively resistant gouge zones of mechanically comminuted and compacted sand grains (Jamison and Stearns, 1982). These narrow zones, which are generally less than 1 mm wide, tend to have lower porosity than the surrounding sandstone. They can generally be regarded as sites of localized strain hardening, but some anastomosing micro-faults do lose cohesion because of weathering or undercutting (Jamison and Stearns, 1982). Similar micro-faulting phenomena have been reported from other highly porous sandstones, including the New Red Sandstone of Scotland. In the New Red Sandstone, these faults occur as closely spaced (<1 m), often conjugate sets, which affect large volumes of the rock (Underhill and Woodcock, 1987). The faults are resistant, and generally stand as small protuberances above the surrounding outcrop. Once again, strain hardening has taken the form of the comminution and compaction of grains, resulting in reduced porosity and increased resistance to surface abrasion.

Deformation of sandstones can also be seen at the mesoscale, although the type of deformation is partly dependent on the mechanical characteristics of the rock. Whereas ductile rocks will tend to attentuate, brittle rock will tend to break or crush. Tricart (1974, Fig. 2.17) provided a striking illustration from the Venezuelan Andes, in which anticlinal folding of quartzitic sandstone has resulted in crushing of beds at the heart of the fold, and in faulting on a flank of the fold. Deformation at this scale is evident in various parts of the Colorado Plateau, especially in association with monoclinal folding. Mechanical crowding of beds within the Tapeats Sandstone is well displayed at the lower, or synclinal hinge of the East Kaibab Monocline exposed in Chuar Canyon (Huntoon, 1989). The consequent crumpling and fracturing of these beds has resulted in very broken outcrops. Less massive or more ductile formations are often highly attenuated. For example, where the dip steepens with increasing depth into the monocline, strata including the Supai Formation sandstones have been stretched out and thinned by 30–60%. These changes are imprinted on the local topography of the sandstones.

Regions of simple post-depositional deformation

In reviewing the influence of large-scale post-depositional deformation on sandstones, we have been mindful of the degree of variation within, as well as between, major groupings of landform assemblages, as was emphasized in the previous chapter. The use of supposed type examples in the comparison of structural or tectonic settings carries an inherent potential for bias just as great as that recognized in the review of climatic influences on sandstone landscapes. Nothing is to be gained by replacing an approach based on an *a priori* climatic determinism with an *a priori* tectonic determinism. Again, therefore, we have looked at a range of examples, this time drawn from areas with differing degrees, styles and ages of deformation, and differing types of sandstones.

Although of great age, vast areas of sandstone on the continental platforms have undergone surprisingly little deformation. For example, the Proterozoic sandstones of the Kimberley and Arnhemland blocks in northern Australia have been, for the main part, subjected only to localized folding and faulting, and still preserve dip planes with inclinations close to those of the original depositional surfaces. Broadly, there are sequences of extensive, cliff-lined plateaus and benches, separated by gentler slopes on more readily erodible rocks; the detailed pattern of topography is very much controlled by lithology and, to a lesser extent, jointing. The effects of deformation are dramatically shown on the southern and eastern margins of the Kimberley Block, where this sedimentary sequence was incorporated in extensive movements along major Precambrian mobile zones. There, the steeply upturned strata have been cut into very long, cliffed escarpments, with often steep backslopes. The underlying shales, which have been excavated by streams cutting back along the strike, form sub-parallel valleys. Along the valley of the Chamberlain River, which flanks the eastern edge of the Kimberley Block, the Pentecost, Warton and King Leopold Sandstones are spectacularly exposed in parallel walls, separated by the Elgee Siltstone and Carson Volcanics, along the Elgee Cliffs which run continuously for about 150 km.

The erosional development of simple faulted scarps in sandstone is well illustrated by the spectacular tower and mesa terrain cut from the gently dipping Paleozoic sequence of southwestern Jordan (Osborn, 1985). The opening of the Red Sea Rift during the Miocene created a major series of normal faults striking northwest, a secondary series of normal faults striking north–northeast, and prominent jointing paralleling the faults. The fall in elevation of the base level of erosion with the opening of the rift unleashed a major, long-lasting period of dominantly linear denudation on the adjacent block. Climates considerably wetter than the present-day desert regime led to the extension of canyons along the faults intersecting the main scarp; and the cutting of numerous tributaries along the dominant joints, progressively isolating masses of sandstone into smaller and smaller compartments. The edge of the block has thus been carved into a complex array of steeply walled towers and mesas that rise up to 800 m above the surrounding pediments and sand flats. Erosion of the generally weakly cemented sandstones has also pushed back the scarp bounding the rift from the position of the master faults.

Tectonic constraints on the denudation of sandstone terrain are well illustrated by the contrasting topography of the Gory Stolwe and nearby Gory Bystrzyckie of southwestern Poland (Latocha and Synowiec, 2008). The Stolwe Mountains have a classic stepped topography on intercalated Cretaceous sandstones, mudstones and marls. There are extensive flat surfaces separated by steep, often cliffed scarps; but the upland is not deeply dissected, and extensive swamps occur in depressions on the summit. The massif has been uplifted 200–300 m along marginal faults, and has been shaped mainly by scarp retreat caused by groundwater sapping and mass failure. In contrast, the Bystrzyckie Mountains, which are built mainly of the same rocks, have fewer extensive structural benches, and are extensively dissected by deeply incised valleys, with long linear slopes. This massif is split by numerous faults that have fragmented the pre-existing topography. Individual fault blocks have not only been displaced, but also tilted. The major valleys are partly the direct result of faulting, but also have been deepened by erosion. In the Stolwe Mountains, deep and extensive groundwater circulation has promoted basal sapping at the expense of surface erosion; but in the Bystrzyckie Mountains, faulting has disrupted the continuity of aquifers, and the exposure of marls and gneiss on the lower slopes has enhanced runoff and erosion.

Monoclinal folding

Because most sandstones are quite brittle, deformation often results in a combination of folding and faulting at the one site. The prominent monoclinal folding and faulting in the Hawkesbury Sandstone on the front of the Blue Mountains west of Sydney is a case in point. As mentioned previously, the Permo-Triassic sediments of the Sydney Basin suffered remarkably little deformation during elevation, with syn-depositional deformation generally exceeding post-depositional deformation. Nonetheless, post-depositional movements along the Lapstone structural complex produced a major topographic break on the east-facing front of the Blue Mountains. The complex consists of a number of related folds and faults trending north–south, with a general southerly plunge. In places, the sandstone is deformed by a single monocline; in others, there is a double monocline; and in some places, a normal or high-angle reverse fault has fractured the rock mass (Branagan and Pedram, 1990). The crest of the structure is broken by a smaller west-facing scarp formed by a series of *en echelon* faults, west-side down. Although the deformation can be attributed mainly to post-depositional movements, probably Cretaceous to Miocene in age (Branagan and Pedram, 1990), its location seems to have been controlled by a much older, deep-seated structure. Gravity patterns suggest the presence of a major escarpment in the folded Paleozoic basement well below the Triassic sequence (Leaman, 1990).

A close link between faulting and monoclinal folding of sandstone is well illustrated also on the Colorado Plateau, where monoclines of Laramide age in the Paleozoic and Mesozoic sedimentary cover formed in response to reverse movements on favourably oriented, pre-existing faults in the Precambrian basement (Huntoon, 1989). In many cases near the Grand Canyon, the initial reactivation of faulting which triggered monoclinal deformation in the overlying sediments continued after the folding had ceased, producing significant displacement of individual beds that is reflected in the topography (Kelly, 1955). There are abrupt stepped offsets at the top of the Precambrian basement; dips increase and the width of folds decrease with depth in the monoclines; and in many places, faulting disrupts the folded rocks on the surface of the monocline. For example, the warped

Fig. 9.2 Hogbacks on an eroded anticline in the Navajo Sandstone near Kayenta, Arizona, USA

surface of the Permian Esplanade Sandstone is broken by two large normal faults on the Lone Mountain Monocline (Huntoon, 1989, Fig. 7–12).

The topographic effect of monoclinal folding on the Colorado Plateau varies with the types and sequence of rocks, the amount of deformation or displacement of the strata, and the degree to which the structure has been breached by erosion (Fig. 9.2). The typical sequence of valley and cuesta development along the strike of eroded monoclines is well illustrated on the Waterpocket Monocline. There, lowlands eroded mainly on the Chinle Formation are backed by a steep escarpment mostly cut in the Wingate, Kayenta and Navajo Sandstones. This sequence is followed by another series of valleys and scarps in the Upper Jurassic and Cretaceous sandstones and shales. The Waterpocket Monocline also shows that the actual sequences of rocks and topography may be more varied than the apparently rigid relationship between the hardness of rock and the position of the valleys and scarps that is presented in many textbook diagrams. Streams quite commonly penetrate through the softer rocks into the underlying sandstones. In places along the Waterpocket Monocline, siltstones and thinly bedded sandstones of the Moenkopi or Chinle formations are exposed on dip slopes or in the footslope of the succeeding scarp; while harder rocks, such as the Entrada Sandstone, are exposed on the floors of strike valleys.

Deeply eroded, gently folded terrain

The topography of the Flinders Ranges of South Australia displays the clear imprint of Proterozoic and Cambrian geosynclinal deposition, of folding mainly of Paleozoic age, and of subsequent deep and episodic denudation (Twidale, 1967, 1971). These ranges are within the Adelaide Geosyncline, a major trough which lies on the eastern margin of the Western Australian shield and into which some 16,000 m of sediments were deposited. Massive sandstones, which are now the main ridge formers, were laid down along the margins of the geosyncline, their distribution shifting with the advance and retreat of the ancient shoreline. The Emeroo Quartzite and the Pound Sandstone are restricted to the western part of the ranges, the ABC Range Sandstone to the eastern part. The structures into which these sandstones, together with limestones and shales of the geosynclinal sequence, were folded are generally simple and open. As a consequence of the simple style of folding and of the great resistance of the ridge-forming rocks, the structural pattern is clearly expressed in the patterns of ridges and valleys, the assemblages of cuestas and hogback ridges, and in the convergence and divergence of ridges associated with plunging folds (Twidale, 1967, 1971). Deep erosion of the folded sequence produced a very marked inversion of topography. The breached anticlines are now the sites of valleys and broad lowlands, such as the 20-km wide Willochra plain in the centre of the ranges; and the synclines are the sites of the main uplands. The most prominent of these synclinal uplands is Wilpena Pound, where the plunging structure of a major syncline is superbly preserved by thick outcrops of very tough Pound Sandstone, some 300 m above the lowlands cut in adjacent anticlines (Fig. 9.3).

Twidale (1967) demonstrated strong structural and lithological controls on hillsides in the sandstone and quartzites of the Flinders Ranges, but also warned that the effects of vertical changes of structure in deeply eroded sequences need to be taken into account. In many places in these ranges, such as The Bluff and Dutchmans Stern, north of Quorn, quartzites and sandstones are cut into a single massive bluff or cliff, beneath which there is a rectilinear debris slope on softer rocks that is broken by minor ledges of resistant beds or lenses of sandstones or quartzite. Many prominent escarpments, such as those of Wilpena Pound and the nearby Elder Range, are compound assemblages consisting of several steep bluffs separated by gentler debris slopes. Joints, rather than bedding, constrain the detailed form of most bluffs and cliffs; but at Buckaringa Gorge, where the ABC Range Quartzite dips at 70–80°, bedding planes are the dominant control of the near-vertical face. In most cases, however, it is on the dip slopes behind the bluffs that the influence of bedding planes is dominant, although the surface of these slopes is often broken by small reverse steep faces that follow joints. Twidale

Fig. 9.3 Wilpena Pound, Flinders Ranges, South Australia. The beds on the side
of the breached anticline stand in steep bluffs above the eroded centre

pointed out that some features which now appear to be discordant with structure, such as streams which cut through sandstones on the nose of plunging anticlines, may well have been once accordant with the structure higher in the stratigraphic section, but have incised from softer to harder rocks as the disposition of strata changed with depth.

The northern Appalachians have been the site for many studies of landforms carved from simple folds. All of these studies have emphasized the close relationship between structural, lithologic and topographic patterns, for the erosion of these dominantly Paleozoic rocks has produced a fine set of ridges and valleys that generally run sub-parallel for considerable distances along the strike, and eventually converge in canoe-shaped forms in the direction of the plunge of the folds. The influence of lithology on topography is very striking indeed; among the main ridge formers, the Pottsville, Pocono and Tuscarora Sandstones are especially prominent. Particularly striking examples of this type of terrain occur in the Shenandoah valley, where thick sandstones preserve the outline of a plunging syncline on Massanuten Mountain, high above the adjacent valley floors carved in fissile shales (Hack, 1965). The relationship between topography and structure is more complex than between topography and lithology. In some places, such as Massanuten Mountain, the anticlines have been breached and the ridges are concentrated on the synclines; but in others, there are anticlinal ridges and synclinal valleys. Ashley (1935) proposed a sequence of structural-relief relationships:

- A gently dipping monocline is higher than a steeply dipping monocline.
- An anticline is higher than a monocline.
- A broad anticline is higher than a narrow anticline.
- A syncline is higher than an anticline.
- A point of juncture of two ridges of a syncline or a breached anticline is higher than either ridge elsewhere.

Ashley's conjecture is worthy of further detailed testing, especially in light of debate about the origins of the accordance of ridge summits, a topic to which we return at the end of this chapter.

Terrain similar to the northern Appalachians, although bearing the imprint of generally steeper dips, occurs in the folded strata of the Macdonnell, Krichauff and James Ranges of central Australia (Mabbutt, 1962). These ranges have a consistent pattern of ridges and narrow valleys which run east–west along the strike of the fold. Many of the ridges lie on steeply dipping and highly resistant sandstones, with the intervening valleys cut in generally narrow bands of softer rocks. The constraints on erosion exerted by the commonly steep dips have produced sub-parallel walls or ribs of sandstones along the centre of the ridges. These ridges are interrupted by rounded slopes on the sides of old valleys perched above the narrow 'water gaps' of the transverse trunk streams that cut across the folded sequence. The central lines of ribs on the ridges are generally flanked by very steeply inclined and faceted hogbacks. Plunging folds are again expressed in the convergence and divergence of ridges.

The Arctic foothill and lowland belts of northern Alaska are characterized by simple, often *en echelon* folding of sequences of Mesozoic sandstones and finer sediments (Chapman and Sable, 1960; Patton and Tailleur, 1964). The folds, which are sub-parallel with an east–west strike, were formed by forces generated by thrust faulting during the Laramide Orogeny in the Brooks Range which lies immediately to the south. Along the margins of the Brooks Range, thrust faulting has partly disrupted the simple patterns of folding. Erosion of the folded sequence has produced a generally rolling terrain broken by long cuesta and whaleback ridges with a local relief of 300–600 m (Fig. 9.4). Erosion of these folds has produced a fine set of anticlinal valleys separated by synclinal mountains and ridges. In the western part of the region, the higher synclinal topography is cut mainly from the sandstones and shales of the Kukpowruk Formation, although the centres of some synclinal mountains are capped by the shales and sandstones of the Corwin Formation. The rocks of these two formations have been stripped from the crests of the anticlines to expose the shales and claystones of the

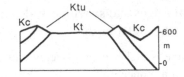

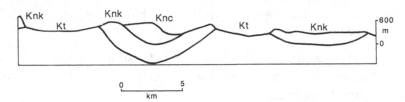

Fig. 9.4 Structure and topography of folded sandstones and shales in the Arctic foothills of Alaska (after Chapman and Sable, 1960; Patton and Tailleur, 1964). Kc Chandler Formation; Ktu Tuktu Formation; Kt Torok Formation; Knk Kukpowruk Formation; Knc Corwin Formation

Torok Formation into which very broad valleys have been cut. The flanks of most of the synclinal prominences, such as Igloo and Poko Mountains, are beautifully benched. This strikingly layered appearance reflects the pronounced inter-bedding of sandstones and shales in the Kukpowruk Formation, the shattering by frost of the numerous bedding planes in this platey sandstone, and the presence of only a thin veneer of periglacial scree on the slopes (Chapman and Sable, 1960). The Corwin Formation has a topography similar to, if more subdued than that of the Kukpowruk, mainly because of the greater proportion of claystone and shales, although the sandstone beds in the Corwin form locally prominent ledges and pinnacles.

The shales and claystones of the Torok Formation again underlie the very broad valleys in the breached anticlines of the eastern part of the region; but the Kukpowruk sandstones are replaced as the main cliff formers by the sandstones of the Tuktu Formation, which form cliffed ridges 300–600 m high (Patton and Tailleur, 1964). The most prominent of these ridges cut in the Tuktu sandstones is the long, 600-m high, south-facing escarpment that rises abruptly from the lowland cut in the Torok shales. The stratigraphic equivalents of the Kukpowruk and the Corwin Formations appear to be the sandstones and shales of the Chandler Formation, which lies above the Tuktu sandstones. The Killik Tongue of the Chandler Formation includes a massive, resistant sandstone which caps the Tuktu escarpment and, further north, forms the prominent hogback ridges that surround the breached core of the Arc Anticline (Patton and Tailleur, 1964). The second major contrast with the western region is the greater outcrop of the

Fortress Mountain Formation which lies beneath, and is exposed south of the Tuktu shales. This formation – which consists of a thick series of greywackes, conglomerates and siltstones – has been carved into hogback ridges and several prominent mesa-like mountains with local relief of up to 600 m. The slopes on the greywackes take the form of long, sometimes rounded ridges, which (in contrast to the slopes on the Tuktu and Kukpowruk sandstones) generally show minimal topographic expression of bedding. On Fortress Mountain, however, a facies change from the sandstones on the north flank to the thick conglomerates on the south flank is matched by a transition from long, relatively smooth slopes to cliffs and large, joint-bounded pinnacles (Patton and Tailleur, 1964). Once again, the highest landforms are on the flanks of synclines, but the folding in this formation is much more disrupted by thrust faulting than that in the younger formations that crop out further from the flanks of the Brooks Range.

An area of simple deformation adjacent to an alpine zone

Although New Zealand is very seismically active, basins on either flank of the axial tectonic belt are not highly deformed. The massive sandstones of the Eocene Brunner Coal Measures, which lie near the northwest coast of the South Island of New Zealand, show much less deformation than the sediments in the adjacent alps. These sandstones form a remarkably bare and rocky surface on the Stockton Plateau where the sequence of deformation is readily interpreted from fracture and fold patterns (Bishop, 1992). Three distinctive series of faults cut the plateau surface:

- Normal faults strike NNE–SSW, one of which – the Kiwi Fault – has displaced the sandstone by some 50 m.
- Reverse faults strike NNW–SSE, and lifted the plateau to its present elevation of 800–900 m during the Neogene.
- Sinistral slip faults strike NW–SE, and display offsets up to 300 m.

During the Late Neogene (Pliocene), WNW–ESE crustal shortening resulted in strong folding against the reverse faults and produced low open folds across the plateau surface. This shortening also produced a strongly developed and steep jointing that cuts most outcrops of the sandstone. In central Otago, Tertiary sandstones that were deposited on an erosional surface cut across schists and greywackes are still preserved on the summits of ranges east of the Southern Alps. Their preservation was, in places, greatly enhanced by the development of diagenetic silcrete up to about 6 m thick, formed by the precipitation of silica from groundwater (Youngson *et al.*, 2005) (Fig. 9.5). The great resistance of these outcrops is demonstrated by [10]Be analysis that indicates maximum ages of

Fig. 9.5 Silcrete boulders in eastern Otago, New Zealand. Growth of anticlinal ridges has disrupted the previously continuous sheets of silcrete

exposure of up to 1400 Ka (Youngson *et al.*, 2005). However, the growth of anticlinal ridges above blind reverse faults during the Quaternary has facilitated erosion of weaker sediment and weathered schist beneath the silcrete. While the silcrete remains relatively undisturbed on ridge crests and gentle back slopes, undercut slabs on the steeper flanks of the ridges are cantilevered and undergo tensile failure.

Complex deformation

Zones of active tectonics

The Himalayas are by far the most outstanding of the young thrust regions of the world, and include some spectacular sandstone and conglomeratic terrain. For example, the massive Muth quartzites, which have an average thickness of 800 m, form many major peaks in the Kumaon Himalayas (Gansser, 1964). But it is the evidence of ongoing thrusting, and of the great variability in the degree of deformation of coarse clastic rocks in the Himalayas, which is of prime concern here. Fracturing and shearing of the massive and cross-bedded sandstones of the Late Tertiary and Pleistocene Siwaliks (Gansser, 1964, 1983), together with the great increase in the elevation of the Himalayas and the Tibetan Plateau over the last few million years, show that thrusting is still active. Quaternary thrusting has displaced and distorted great thicknesses of the Siwaliks in Bhutan, giving a marked regularity of northerly dips of 50–60° and localized overturning of beds (Gansser, 1983). The thrusting of several imbricated slices of the Siwaliks in Bhutan has produced a series of prominent ESE–NNW trending ridges and hog-backs. There are subsequent strike–pull-apart structures, high localized bulges, small sedimentary basins and indentation of the ridges along the thrust fronts (Delcaillau *et al.*, 1987). These structures are reflected in the complex morphology of the scarp-forming sandstones.

Gansser also reported consistently northerly dips in older quartzites on ridges near the Chu valley in Bhutan, and added that these rocks are highly cleaved and generally show much local disturbance. Yet the degree of distortion decreases rapidly away from the thrust faults, and some very large sheets of sediment have been uplifted to great elevations without being markedly distorted. For example, the 6700-m peak of Kailas in southern Tibet is a 'surprising and beautiful remnant' of a vast sheet of undisturbed, near-horizontal detrital material that was about 4000 m thick (Gansser, 1964, p. 10). The horizontal beds in the bulk of the Kailas sediments can be traced southward until they suddenly become distorted near the margin of the Himalayan thrust. Distortion also varies with the rigidity of the rocks. The thinner intercalated sandstones show intense disharmonic folding at some distance from the thrust, while the coarser Kailas beds are still undisturbed (Gansser, 1964).

The mountains of Taiwan are also the product of active thrusting, but, in contrast to the continental collision that elevated the Himalayas, they are the focal point of continental and island arc collision (Suppe, 1987). Miocene and Paleogene (Early Tertiary) sediments have been piled up and thrust over Pleistocene sediments in the western Taiwan foredeep. Suppe likened the form of the thrusting to a compressive wedge in front of a tectonic 'bulldozer'. The effects of active mountain building in this island are exemplified by the 300-m high cliffs cut in the Upper Miocene sandstone of the Nanchuang Formation, which has been tilted to 15–20° while being raised to an elevation of 2500 m. These cliffs consist of extremely bold faces, disrupted by minor benches, and are being dissected by stream erosion. Not only are these mountains being raised rapidly, they are also being eroded rapidly, with denudation rates estimated at between 5–9 km/Ma (Li, 1975).

The greywacke terrain of New Zealand is also a highly mobile tectonic zone. Paleozoic and Mesozoic greywackes were intensely folded in the Jurassic and Early Cretaceous, and then planated in the Early Tertiary. In the Neogene (about 25 Ma), mountain building was renewed, with large-scale faulting. Collision of the Australian and Pacific Plates has resulted in subduction in oceanic trenches to the northeast and southwest of New Zealand, but tectonic stresses generated in the buoyant continental rocks have resulted in strike–slip shearing at a rate of about 40 km/Ma. These movements have been accompanied by rapid uplift, with rates of 8–10 km/Ma along the western flank, and approximately 1 km/Ma on the eastern flank, of the Southern Alps (Tippet *et al.*, 1995; Coates, 2002). Erosion has kept pace with uplift, as a 20-km thickness of rock – amounting to about 500,000 km^3 – has been stripped during the last 5 million years (Coates, 2002). Rapid uplift and erosion has also occurred along the axial tectonic belt on the North Island (Campbell and Hutching, 2007).

The close fracturing and generally vertical dip of the greywacke has facilitated the development of an intensely dissected, 'feral' landscape (Suggate, 1982) (Fig. 9.6). Many major streams closely follow the trace of fault lines; tributary streams are closely spaced; slopes are steep; mass failure is common; and extensive scree slopes are widely developed (Crozier *et al.*, 1982). Such intensely dissected terrain is particularly well developed in the Kaikoura Ranges in the northeastern part of the South Island. Further south, many of the main valleys were widened by glacial erosion, and the outwash on their floors was shaped into extensive terraces by post-glacial uplift and erosion. The Waimakariri River is a case in point, although it has cut a narrow gorge where it flows through the Torlesse Range onto the Canterbury Plains.

Similar rugged but lower landscapes occur in the greywacke of the North Island. The southern end of the island consists of a series of tilted blocks, each uplifted on the western side of parallel strike–slip faults. Lowlands, including Wellington Harbour, occupy fault depressions that have steep scarps on their west flanks and inclined back slopes on their east flanks. The main streams have extended along the faults, and tributary streams have intensely dissected the tilted blocks. Frictional resistance to movement along the dominant NE–SW striking faults has created a series of secondary N–S faults, many of which have also been exploited by stream erosion. The degree of dissection becomes even more pronounced further north in the Tararua Range, which is bounded on either side by faults, and is rent by faults extending along its axis. Continued movement along the faults is demonstrated by the offset ends of ridges, by abrupt changes in the orientation of stream channels, and by small scarps that cut across alluvial terraces (Campbell and Hutching, 2007). The recent uplift of the block is also demonstrated by the antecedent course of the Manawatu River, which flows through a narrow gorge from the fault depression on the east side to the lowland on the west side (Fig. 9.7).

Regions of Caledonide orogeny

Britain and Greenland bear the strong imprint of thrusting that occurred during the Caledonide orogeny of Siluro-Devonian times (c. 420–350 Ma). Sections of

Fig. 9.6 The Torlesse Range, South Island, New Zealand. Rock avalanches flow from the greywacke peaks of the Southern Alps

Fig. 9.7 The Manawatu River has cut its gorge through greywackes of the Tararua Range, North Island, New Zealand

the Torridonian landscape of northwestern Scotland described in the previous chapter have been displaced by the Moine Thrust. For example, the summit of Meall a Ghuibhais, near Kinlochewe, consists of a mass of Torridonian Sandstone thrust over stratigraphically higher limestones, shales and grits; these in turn lie on the *in situ* Torridonian which forms the lower slopes of the mountain. The mountain is asymmetric because of this rearrangement of the rocks by thrusting. The steeper northwestern side is cut mainly in the Torridonian Sandstone, the upper part of which is displaced and the lower part of which is still *in situ*. On the southeastern side, however, a long, gentler slope has been cut across the sediments over which the thrust originally passed.

Thrusting has also influenced the carving of landforms from the Cambrian and Devonian sandstones of the Caledonides in southern and western Norway. The

Cambrian *sparagmites* of southern Norway are dominantly arkosic sandstones and conglomerates, with greywackes and, especially at the top of the sequence, quartz sandstone or quartzite (Holtedahl, 1960; Strand and Kulling, 1972). Where dips are gentle, a simple benched topography has developed; but thrusting and folding have produced complex arrays of rock that are reflected in a more complex topography. The relatively rigid *sparagmite* series are not easily folded, but have been displaced by thrusting over fissile shales. The resulting *nappes*, or displaced sheets of sandstone, are themselves thrust along secondary faults, forming imbricated structures (*schuppen*). The imbricated thrust sheet on the western side of Lake Mjosa, for example, has been carved into a series of sub-dued steps, with breaks of slope mainly along the imbrication planes (Holtedahl, 1960). But the main breaks of slope in the strongly benched topography of the thrust sheet at Lake Femunden appear to be controlled more by the lithological contrasts within the *sparagmites* and conglomerates than they are by the imbrication planes (Holtedahl, 1960, Fig. 55).

The topographic effects of thrusting are also obvious in the Devonian sand-stones of western Norway. In the eastern Kvamshesten district, for example, a displaced mass of sandstone and conglomerate rises in vertical cliffs above the much gentler slopes developed on the schists below the thrust fault (Holtedahl, 1960, Fig. 96). Cliffs occur where Devonian sediments overlie metamorphic rocks in parts of the Solund islands; but there it is the Devonian sediments which have been over-ridden by gabbros displaced along thrust faults, and it is the gabbros which dominate the topography (Holtedahl, 1960). Pleistocene glaciation has modified the topography here also. The steep, in places almost conical peaks of the Rondane, into which cirques have bitten deeply (see Fig. 2.10), are rem-iniscent of the peaks of the Torridonian Sandstones of Scotland (although the Rondane rise above the main upland surface rather than a coastal lowland).

A visually striking landscape has been carved from sandstones extensively contorted and displaced by folding and thrusting in the Caledonides of eastern Greenland. Yet the topography does not reflect the complexity of the geology. The superb photographic presentation by Haller (1971) of the magnificent sea and glacial cliffs of this region provides a ready means of relating landforms to structure and lithology. The Caledonian orogeny resulted in the disruption of large areas of Precambrian rocks, especially of the Quartzite Series in which the dominant rocks are orthoquartzites. Haller's photographs show that pronounced bedding and substantial differentiation in the rocks of the Quartzite Series exert surprisingly little control on slope form. The tightly folded beds in the Caledonian synclinal structure on southern Andree Land, and the steeply dipping beds on Ymer Island above Kejser Franz Joseph Fjord (Haller's Photos 31 and 32) are cut by precipitous slopes that display only minor ledges and cliffs controlled by particularly massive beds. Even where these strata have been only slightly tilted, as in parts of Scoresby Land (Haller's Photo 29), they have been cut into long narrow ridges that display only minor bedding control.

The extensive Devonian sandstones of eastern Greenland have been contorted and displaced to varying degrees. Near Kejser Franz Joseph Fjord, the three major units of the Devonian molasse – namely the Cape Kolthof Series, the Cape Graah Series and the Mount Celsius Series, in which sandstones are dominant – show little evidence of deformation apart from normal faulting and broad warping. These rocks, especially the massive beds at the base of the Cape Graah series, form extremely steep, basically uniform walls along the fjord (Haller's Photo 92). On the plateau surface, however, benching is prominent. Sections through Sederholms Berg on the Gauss Peninsula show a steeply dipping faulted contact between contorted Cape Kolthof beds and uniformly dipping Cape Graah beds, above which are flat-lying Mount Celsius beds (Haller, 1971, Fig. 101). Irregular slopes on the lowest beds contrast to the tabular forms on the uppermost beds. The Old Red Sandstone is also deformed to varying degrees. At Dusens Fjord, where it is tilted without much internal disruption, bedding seems to exert only a secondary influence on topography, as is also the case at Whittards Bay in northern Hudson Land (Haller's Photos 88 and 110). At the latter site, the Old Red Sandstone is contorted, dips very steeply, and has been over-ridden by the Limestone and Dolomite Series. Nonetheless, the precipitous slopes carved by ice extend from one rock type to the other without any major change in form. The reason for the general suppression of lithological and bedding constraints in the east Greenland landscape probably results from a combination of factors, including fracturing during orogenesis, oversteepening of slopes due to deepening of the fjords by glacial erosion, and to shattering of outcrop in this extreme climate.

Regions of Laramide orogeny

In the Canadian Rockies, the Laramide Orogeny of Cretaceous to Late Paleocene times involved large-scale thrust faulting from the west which resulted in 200–300 km of crustal shortening. The compressional regime was followed by crustal stretching involving normal faulting and, in the west, strike–slip faulting. As a consequence, vast masses of rock were broken and displaced, and the Rockies display fine examples of thrusting in sandstones and quartzites. Thrusting in the main ranges of the central Canadian Rockies displaced Precambrian rocks, essentially as large single sheets, for distances of about 40 km. These rocks included some of the most resistant in the Rockies, such as the quartzites and conglomerates of the Gog and Atan Groups. The very tough Gog quartzites have been thrust to high elevation, and form enormous cliffs and

Fig. 9.8 Highly fractured Aztec Sandstone, The Valley of Fire, Nevada, USA. Excavation of the numerous fracture planes has produced an extremely irregular terrain with rounded pinnacles

summits on some of the largest mountains, such as Mount Edith Cavell, Pyramid Mountain and other major peaks near Jasper (Gadd, 1986). Displaced and fractured masses of Paleozoic sandstones and quartzites are also widespread. The Early Ordovician Monkman and Tipperary Quartzites are major cliff formers in the eastern parts of the main ranges; and the Mount Wilson Quartzite forms very prominent cliffs near the summit of the peak after which this formation is named. Cretaceous sandstones and conglomerates of the Cadomin Formation form resistant cappings on ridges in the western foothills of the central Rockies, and, where steeply tilted, form prominent hogbacks (Gadd, 1986).

Deformation, intense fracturing and discontinuity of outcrop due to thrust faulting is well illustrated by the Jurassic Aztec Sandstone in the Valley of Fire National Monument of southern Nevada (Fig. 9.8). A thick sequence of Paleozoic carbonates was thrust across, partly incorporated and also deformed Mesozoic sediments which included the Aztec Sandstone (Longwell, 1949). The main outcrop of the sandstone is separated from other smaller exposures by faulted blocks or by thrust sheets. The degree of deformation of the Aztec Sandstone varies considerably. The bedding generally dips at 20–30°, but more steeply (60–70°) near the margins of the thrust sheets; and adjacent to the small Summit

Thrust, beds of the sandstone are overturned. The beds are displaced along several major and numerous minor normal faults; and a portion of the sandstone has been incorporated in the thrust faulting, and now forms a klippe overlying Cretaceous sediments immediately to the east of the main Aztec outcrop. Numerous outcrops of the sandstone are split by intense jointing and small faults which having been picked out and accentuated by erosion to form a very rugged, sometimes needle-like, topography.

The effects of earthquakes

Mass failure triggered by earthquakes is extremely important in shaping the landforms of tectonically active zones. A survey of 40 historical earthquakes indicated that major seismic shocks may trigger landslides over an area of $500,000 \, \text{km}^2$ (Keefer, 1984). During an earthquake in the Torricelli Mountains of New Guinea in 1935, a 12-cm thick layer of weathered granite, and a 7.4-cm thick layer of mixed sedimentary rocks, including sandstone, were stripped from an area of $238 \, \text{km}^2$ by a shock of about 7 on the Richter scale (Simonett, 1967). Given the probable return period of seismic events of about this magnitude, together with the effect of minor shocks coinciding with high rainfalls, Simonett estimated that denudation from landslides triggered by earthquakes in this region would amount to about 10 cm every 70–100 years. As Löffler (1977) concluded, mass movement induced by earthquakes must be regarded as a major factor in the topographic development of tectonically mobile places like New Guinea. Studies of major failures elsewhere also indicate that seismic shock is an important denudational agent in many sandstone landscapes, and not just those in contemporary orogenic zones.

Seismic triggering of mass movement in the greywackes of New Zealand has been well documented. In a $10,000 \, \text{km}^2$ area in the central part of the Southern Alps, 46 rock avalanches ranging in age from 21–10, 250 years have been dated by ^{14}C and the thickness of weathering rinds (Whitehouse, 1983). A marked concentration of avalanches in the period 1400–1500 A.D. indicates that there was a regional seismic event that caused widespread and massive slope failure (Orwin, 1998). The largest recent failure was triggered by the Arthurs Pass earthquake in 1929, when some 60 million cubic metres collapsed from Falling Mountain in the Southern Alps. However, seismic effects can be complex, as is illustrated by a major failure in the eastern part of the North Island (Pettinga, 1987). The Ponui landslide was a very large wedge failure along bedding planes in mudstones and along associated fractures in overlying massive sandstone strata. The failure was triggered by the 1931 Napier earthquake which led to an initial 3 m movement. Subsequent movement was not detected before large-scale failure occurred in 1976, which incorporated an entire sandstone ridge 1100 m long and which has left a 60-m high scarp in sandstones. The sliding plane, which dips at 10–36°, parallels bedding in a succession of porous sandstones and weak, slake-prone mudstones. This surface has a thin, highly plastic coating of gouge that seems to be an extension of a tectonically formed shear zone in a nearby thrust fault. Residual effects such as these need to be considered when assessing the significance of earthquakes.

Seismic triggering of very large failures has also been well documented from the North American cordillera, including several in sandstones. The Great Alaskan earthquake of 1964 caused a great mass of greywacke and argillite to collapse onto the adjacent Sherman Glacier (McSaveney, 1978). The source of the avalanche was in a steeply dipping, extensively jointed, tabular block of hard, coarse-grained greywacke and hard, non-fissile argillite. The shaking apparently triggered failure along a pre-existing weakness, for the avalanche scar follows a set of intersecting, hematite-coated, curved fault surfaces that closely parallel the steeply dipping bedding. The Gros Ventre slide of Wyoming incorporated massive cross-bedded sandstones and quartzites, overlying a sequence of shales and weak sandstones in which the failure plane was seated (Voight, 1978). It occurred in a seismically active zone, but was apparently triggered by a relatively minor earthquake – with an intensity of about 4 on the Mercalli scale – that took place the previous night. Heavy rain and melting snow, which had saturated the mass and greatly reduced its strength, enhanced the effect of this minor shock.

Although relict rock falls in the arid sandstone terrain of southern Jordan have been attributed to wetter climates in the past, cosmogenic dating indicates that some of them may be the result of seismic shock generated by movement on the nearby Dead Sea fault (Matmon *et al.*, 2005). Three discrete rock fall events occurred at 31 Ka, 15 Ka and 4 Ka. Cliff retreat, calculated by the timing and estimated thickness of material removed during these events, ranges between 0.14–2.0 m/1000 years. As these rates are an order of magnitude greater than the maximum rates of weathering measured at Petra (Fig. 9.9) by Paradise (2003), infrequent events such as these seem to play the main role in the denudation of the cliffs of this region.

Even in tectonically quiescent regions, there may be topographic expression of past earthquake activity. Extensive relic avalanche deposits mantle slopes beneath cliffs of Hawkesbury Sandstone in the southern Sydney Basin. These deposits, which consist of large boulders supported by a sandy-clay matrix, are generally much coarser, and certainly travelled much further than almost all contemporary avalanches (A. R. M. Young, 1977). The only contemporary failure of near-comparable size was triggered by subsidence following coal extraction. Although these prehistoric failures were previously thought to have occurred in times of higher rainfall, a climatic origin seems less likely now that

Fig. 9.9 The city of Petra was carved into the sandstone mountains of Jordan by the Nabateans from the fifth century BCE onwards, and later modified under Roman occupation. (Photo: P. Migon)

it is known that the required magnitude of rainfall would have to have been vastly greater than the extreme events of the present day. This fact became apparent in 1984 when a prolonged period of rainfall, which largely saturated the slopes, was followed by a downpour with an estimated return period more than 200 years. Despite these extreme conditions only shallow, short-distance avalanching orders of magnitude smaller than the prehistoric failures occurred. Although this region has been commonly regarded as seismically quiescent, it in fact lies in an active seismic zone in which shocks of about 6 on the Richter scale have been recorded on several occasions. We suggest that the great contrast between contemporary and ancient avalanching reflects a significant scaling down of the seismic regime.

Erosion as a cause of deformation

So far we have considered deformation of sandstone as a fundamental constraint on the subsequent or, in some cases contemporaneous, erosional shaping of the land surface. The relationship is not, however, invariably one way; erosion can be the cause of, or at least an important contributing factor to, the deformation of sandstone.

The relationship between erosion and gravity tectonics in sandstone terrain can be seen in the numerous, large rockslides in the Dakota Group along the Front Range of Colorado (Braddock, 1978). Valleys cut along the flanks of hogbacks along the Front Range reveal that a sequence of sandstone and clayey siltstones have been sheared near the base of the hogbacks, and have moved down-dip as coherent, detached blocks over distances of 300–1250 m. The growth in the height of a hogback, resulting from erosion of the tilted sequence, causes a progressive disparity between maximum and minimum stresses in the siltstones near the base of the slope. Rising porewater pressures at the base of the slope, together with creep in the clayey siltstones, may contribute to failure. The failure extends down-dip along bedding planes in the siltstone until the ridge crest of dipping sandstone intersects the valley floor. At that point, a transgressive fault – which shears upwards through the sandstone – develops, and the detached slab moves up it to ride over the land surface. The displaced blocks range in width from about 270 m to 4.5 km. Braddock emphasized that, as these failures are dependent on critical erosional and rock strength thresholds, they are an ongoing phenomenon in this tilted sequence. He estimated that they range in age from about 30 Ma, when the sequence began to be eroded, to only a few thousand years.

Erosion is apparently also an important factor in the mechanics of shallow thrust faults. By modifying the thickness of the advancing sheet, it is a major determinant of the overburden stress, and hence of the basal resistance to movement and the maximum stable length of the over-riding slab of the thrust. The thrust will continue to advance as an essentially coherent slab until basal resistance exceeds the strength of the slab. Thereafter the slab breaks; imbrication and tectonic overlap occur; and the thrust ceases to advance (Willemin, 1984). These relationships are well illustrated in the truly astonishing instance of the Keystone Thrust of Nevada, in which a mountain more than 4 km high advanced 25–50 km across a land surface cut in sandstone (Johnson, 1981). That the displaced mass did indeed move across the land surface is strikingly demonstrated by the stream channels cut into the surface of the Jurassic Aztec Sandstone. These channels which are now exposed beneath the thrust plane under the displaced carbonate mass contain gravels derived from that mass, thereby heralding its approach (Fig. 9.10). The initial shearing incorporated surface layers of the underlying sandstone. Then, as the strength of the sandstone was increased by induration in response to compression, the movement occurred along a thin gouge on the thrust plane. Willemin (1984) estimated that erosion of the mass to about 2 km as it advanced allowed the horizontal force of about 2000 MPa to push the still coherent mass some 38–58 km at rates of probably 4–8 km/Ma.

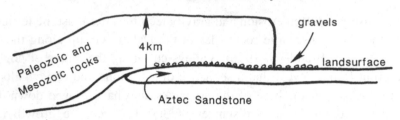

Fig. 9.10 Development of the Keystone Thrust, Nevada (after Johnson, 1981)

Planation

The degree to which the structural and lithological patterns described thus far account for variations in the topography of sandstones remains problematic, especially in the light of the continuing debate about the efficacy of planation. Can accordant summits of places like the northern Appalachians be explained in terms of rock resistance and the spacing of streams, or are they remnants of planation surfaces which cut across a wide variety of rocks? The scheme of cyclical planation proposed by Davis received much criticism, most penetratingly from Hettner (1928), but in a more widely known version from Hack (1960, 1965), and it is rarely invoked today. Hack's rejection of Davis' cyclical interpretation of the Appalachian landscape, and his alternate proposal of a scheme in which topography is in dynamic equilibrium with erosional forces and rock resistance, have been very widely, but not universally accepted. Results of cosmogenic dating of sandstones on high summits in the Appalachians are at odds with Hack's assertion that ridge crests and valley floors are being lowered at approximately the same rate. Whereas valleys in the region are being incised at 30–180 m/Ma, dating by ^{10}Be indicates that sandstone and conglomerate of the Pottsville Groups on Dolly Sods, an undulating upland in West Virginia, are being lowered at only 5.7 ± 2.5 m/Ma (Hancock and Kirwan, 2007). Fairbridge (1988) had earlier argued that, although the concept of planation has at times been carried to unreasonable extremes, the accordant ridges of the Appalachians are indeed remnants of ancient planation surfaces. In reaffirming the broad outline of Davis' interpretation, Fairbridge pointed out that the quartzites on which the bevelled higher ridges lie change with increasing depth to very porous quartz-sandstones. He attributed the upper quartzitic zone to the development of very thick silcrete on and below a planation surface which was slowly being elevated. Be that as it may, planation of even the hardest quartzitic rocks can be clearly demonstrated from other parts of the world.

   Mabbutt (1962) has demonstrated the bevelling of steeply dipping sandstones and quartzites in the Macdonnell and Davenport-Murchison Ranges of central

Fig. 9.11 Spectacular truncation of steeply dipping Proterozoic sediments in the Osmond Range, Kimberley region, northwestern Australia

Australia. Twidale (1983) has described numerous instances of bedrock structures truncated by topographic surfaces, including the surface cutting across the prominent sandstone cliffs which are developed in monoclinally folded sandstones on the walls of Wittenoom Gorge in the northern part of Western Australia. In the East Kimberley region, a few kilometres north of the Bungle Bungle Range, vertically dipping Proterozoic sandstones are truncated by a horizontal knife-edged cut which can be traced northwards across the summits of cuestas in younger Proterozoic sandstones (Fig. 9.11). The much less resistant and generally highly dissected greywackes of the tectonically mobile North Island of New Zealand also bear remnants of a major planation surface that once truncated them (Crozier *et al.*, 1982). A similar well-preserved surface in the South Island, that was previously known as the Otago Peneplain and is now termed the Waipounamu Surface, has been shown to have had a complex development involving both fluvial and marine erosion (Campbell and Hutchings, 2007). The highly contorted and thrusted Paleozoic and Mesozoic sandstones and shales of the foothills immediately adjacent to the Brooks Range in Alaska are also cut by an extensive erosional surface. The degree to which the contorted sediments of this region are planated, and especially the extremely sharp boundary of the planation surface at the foot of the main frontal thrust in limestones of the Brooks Range, are truly remarkable

(Patton and Tailleur, 1964). These examples could be repeated many times over, but they suffice to make the point that widespread planation has been a major factor in the development of many sandstone landscapes.

This is not to say that planation surfaces necessarily developed in the manner specified or at the rate implied in the Davisian model. R. W. Young (1977a) has argued that surfaces cut across sandstones and argillaceous rocks in the upper part of the Shoalhaven catchment in southeastern Australia were graded to structural knickpoints far above the coastal base level, and that they expanded at a very slow rate indeed (R. W. Young, 1983c; Young and McDougall, 1985). Other surfaces of similar extent, such as the Cannonville surface (Gregory and Moore, 1931), which clearly truncate strata on parts of the Colorado Plateau, also may have originated in this manner.

Structural and lithological controls are still evident even on extensive planation surfaces. The Palaeic surface of the Varanger Peninsula in northern Norway is a case in point (Fjellanger and Sorbel, 2007). Although this peninsula is visually dominated by a wide, undulating skyline, detailed mapping has revealed three distinct levels:

- An upper inselberg level lies at 500–700 m, and correlation with erosional boundaries between depositional formations in the adjacent Barents Sea suggests a Mesozoic age to this level.
- The main level lies at 300–500 m, and is considered to be of early Tertiary age.
- The lower level lies at 150–300 m and is incised by the main fjords.

In general, the upper level is dominated by quartzites, the main level by sandstones, and the lower level by shales and mudstones. However, all three levels truncate a variety of rocks. Bedding controls are obvious on massive sandstones in the northeast. Relatively flat surfaces have developed where the quartzites are gently dipping; but steep ridges and escarpments occur where the quartzites dip steeply. For example, an exhumed Proterozoic structure, which dips at 40–50° in the quartzites, coincides with a steep escarpment that rises 250 m above the main level.

Tricart and Cailleux (1972), Büdel (1982) and many other European geomorphologists have argued that planation surfaces are essentially diagnostic of tropical humid climates which produce a highly weathered surface layer that facilitates erosion of broad, pediment-like valleys, rather than narrow incised ones. For example, Busche (1980) interpreted the planation surface that transects the dip of the Nubian Sandstone in central Libya as a Cainozoic etchplain, formed under tropical climates that were at least seasonally humid. In some cases, such as this one, the conjecture is apparently supported by evidence of relic lateritic weathering (although we have argued earlier that such weathering does not require tropical climates). It must be pointed out, however, that the examples of tropical surfaces on which this climatic hypothesis of planation surfaces is based are almost invariably from tropical cratons. They certainly are not typical of mobile areas of the tropics like New Guinea, where greywackes and arenaceous sandstones, along with most other types of rocks, are deeply and intensely dissected by narrow V-shaped valleys (Löffler, 1977). Some planation surfaces may well have formed under tropical climates, but there is no reason to believe that all did. Planation may be more the result of tectonic stability, than of a particular climatic regime.

Conclusions

The great increase of interest in sandstone landforms since 1992, when the predecessor of this book appeared, is highly commendable. A major gap in geomorphology is being successfully closed, so that the study of these landforms is now comparable to those of the landforms in limestone and granite behind which it once lagged so far.

Yet major tasks remain to be addressed:

- Although many important new studies have appeared, especially from South America, Australia, central Europe and China, little seems to be known of sandstone landforms over large areas of the world. Some areas have not been studied in any detail, while the findings of others are still to appear in forms readily accessible for the international audience.
- The increasing attention given to solutional weathering in quartz sandstone, a subject still virtually in its infancy in 1992, is heartening. But the widespread neglect of the implications for sandstone geomorphology of findings from rock mechanics and engineering geology is disappointing. It is especially so because substantial foundations had already been laid by Scheidegger, by Selby and his co-workers, and by the review of the subject outlined by us. Of the papers in a recently published collection of studies on sandstone landforms, only one dealt in any detail with the mechanical properties of the rock. Unless this shortcoming is addressed, quantity may be mistaken for quality of research.

As the development of any assembly of sandstone landforms is contingent on the intersection of multiple causal chains of events, we again appeal for flexibility of analysis rather than adherence to some rigid theoretical framework. But we also emphasize the need to 'map', as Bateson put it, our observations onto fundamental scientific concepts rather than on popular heuristic devices. Finally, we hope that this revision and extension of the earlier book conveys some measure of the longstanding interest that sandstone landscapes have held for us.

References

Adamovic, J. (2005). Sandstone cementation and its geomorphic and hydraulic implications. *Sandstone Landscapes in Europe – Past, Present and Future. Proceedings of the 2nd Conference on Sandstone Landscapes, 25–28.05.2005*, Vianden (Luxembourg), Ferrantia 44, 249 p.

Adamovic, J. and Kidston, J. (2008). Sandstones and their attributes. In *Sandstone Landscapes*, eds. H. Hartel, V. Cilek, T. Herben, A. Jackson and R. Williams. Academia, pp. 13–24.

Adamson, D., Selkirk, P. M. and Mitchell, P. (1983). The role of fire and lyrebirds in the sandstone landscape of the Sydney region. In *Aspects of Australian Sandstone Landscapes*, eds. R. W. Young and G. C. Nanson. Wollongong: Australian and New Zealand Geomorphology Group, pp. 81–93.

Ahnert, F. (1960). The influence of Pleistocene climates upon the morphology of cuesta scarps on the Colorado Plateau. *Annals of the Association of American Geographers*, **50**, 139–156.

Aldrick, J. M. and Wilson, P. L. (1990). *Land systems of the southern Gulf region, Northern Territory, Technical Report 42*: Conservation Commission Northern Territory, Australia.

Alexandrowicz, Z. (1989). Evolution and weathering of pits on sandstone tors in the Polish Carpathians. *Zeitschrift für Geomorphologie*, **33**, 275–289.

Alexandrowicz, Z. (2006). Distribution of sandstone tors controlled by geological patterns in the Polish Carpathian Foothills. *Abstracts of the 9th International Symposium on Pseudokarst*, 24–26 May 2006, Bartkowa, Poland, Institute of Nature Conservation, Polish Academy of Sciences, Cracow.

Alexandrowicz, Z. and Urban, J. (2005). Sandstone regions of Poland – geomorphological types, scientific importance and problems of protection. *Sandstone Landscapes in Europe – Past, Present and Future. Proceedings of the 2nd Conference on Sandstone Landscapes, 25–28.05.2005*, Vianden (Luxembourg), Ferrantia 44, 249 p.

Allen, J. R. L. (1971). Transverse erosional marks of mud and rock: their physical basis and geological significance. *Sedimentary Geology*, **5**, 167–385.

Andre, M. -F. (1997). Holocene rockwall retreat in Svalbard: a triple-rate evolution. *Earth Surface Processes and Landforms*, **22**, 423–440.

Andrews, P. B., Bishop, D. G., Bradshaw, J. D. and Warren, G. (1974). Geology of the Lord Range, central Southern Alps, New Zealand. *New Zealand Journal of Geology and Geophysics*, **17**, 271–299.

Ashley, G. H. (1935). Studies in Appalachian Mountain sculpture. *Geological Society of America Bulletin*, **46**, 1395–1436.

Aston, S. R. (1983). Natural water and atmospheric chemistry of silicon. In *Silicon Geochemistry and Biogeochemistry*, ed. S. R. Aston. London: Academic Press, pp. 77–100.

Aubrecht, R., Brewer-Carías, C., Šmída, B., Audy, M. and Kovácik, L. (2008). Anatomy of biologically mediated opal speleothems in the world's largest sandstone cave Cueva Charles Brewer, Chimantá Plateau, Venezuela. *Sedimentary Geology*, **203**, 181–195.

Aucamp, J. P. and Swart, D. P. R. (1991). The underground movement in Zimbabwe. *Bulletin of the South African Speleological Association*, **32**, 79–91.

Audy, M. and Šmída, B. (n.d.). 'Speleologie v kvarcitovch masivech'. Retrieved 4 September 2007, from http://audy.speleo.cz/quarzit/index.html.

Augustinus, P. C. (1992). The influence of rock mass strength on glacial valley cross-profile morphometry. *Earth Surface Processes and Landforms*, **17**, 391–451.

Augustinus, P. C. and Selby, M. J. (1990). Rock slope development in McMurdo Oasis, Antarctica, and implications for interpretations of glacial history. *Geografisker Annaler*, **72A**, 55–62.

Baker, V. R. and Pickup, G. (1987). Flood geomorphology of the Katherine Gorge, Northern Territory, Australia. *Geological Society of America Bulletin*, **98**, 635–646.

Barton, N. R. and Choubey, V. (1977). The shear strength of rock joints in theory and practice. *Rock Mechanics*, **10**, 1–54.

Bateson, G. (1973). The science of mind and order. In *Steps to an Ecology of Mind*, ed. G. Bateson. London: Granada, pp. 21–31.

Battiau-Queney, Y. (1984). The pre-glacial evolution of Wales. *Earth Surface Processes and Landforms*, **9**, 229–252.

Beckwith, R. S. and Reeve, R. (1969). Dissolution and deposition of mono-silicic acid in suspensions of ground quartz. *Geochimica et Cosmochimica Acta*, **33**, 745–750.

Beloussov, V. V. (1980). *Geotectonics*. Berlin Springer.

Bennett, P. C. (1991). Quartz dissolution in organic-rich aqueous solutions. *Geochimica et Cosmochimica Acta*, **55**, 1781–1797.

Bennett, P. C., Melcer, M. E., Siegel, D. I. and Hassett, J. P. (1988). The dissolution of quartz in dilute aqueous solutions of organic acids at 25 °C. *Geochimica et Cosmochimica Acta*, **52**, 1521–1530.

Berner, R. A. (1978). Rate control of mineral dissolution under earth surface conditions. *American Journal of Science*, **278**, 1235–1252.

Besler, H. (1985). Reliefgenese in den immerfeuchten Tropen (Kalimantan Timur/ Borneo). *Zeitschrift für Geomorphologie Supplement Band*, **56**, 13–30.

Beuf, S., Biju-Duval, B., Charpal, O., Rognon, P., Gariel, O. and Bennacef, A. (1971). *Les Gres du Paleozoique Inferieur au Sahara*. Paris: Editions Technip, Publications de l'Institut Francais du Petrole.

Bird, M. I. and Chivas, A. R. (1988). Stable-isotope evidence for low-temperature kaolinitic weathering and post-formational hydrogen-isotope exchange in Permian kaolinites. *Chemical Geology (Isotope Geoscience Section)*, **72**, 249–265.

Bird, M. I. and Chivas, A. R. (1993). Geomorphic and palaeoclimatic implications of an oxygen-isotope chronology for Australian deeply weathered profiles. *Australian Journal of Earth Sciences*, **40**, 345–358.

Bird, M. I. and Chivas, A. R. (1995). Palaeoclimate from Gondwanaland Clays. *Proceedings of the 10th International Clay Conference*, Adelaide.

Bishop, D. J. (1992). Neogene deformation in part of the Buller Coalfield, Westland, South Island, New Zealand. *New Zealand Journal of Geology and Geophysics*, **35**, 249–258.

Bjelland, T. and Thorseth, I. H. (2002). Comparative studies of the lichen–rock interface of four lichens in Vingen, western Norway. *Chemical Geology*, **192**(1), 81–98.

Blackwelder, E. (1929). Cavernous rock surfaces of the desert. *American Journal of Science*, **17**, 393–399.

Blair, R. W. (1984). Development of natural sandstone arches in southeastern Utah. *Proceedings of the 2nd Conference of the Australian and New Zealand Geomorphology Group*.

Blong, R. J., Riley, S. J. and Crozier, R. J. (1982). Sediment yield from runoff following bushfire near Narrabeen Lagoon, NSW. *Search*, **13**, 36–38.

Blume, H. and Remmele, G. (1989). A comparison of Bunter Sandstone scarps in the Black Forest and the Vosges. *Catena (Supplement)*, **15**, 229–242.

Bögli, A. (1960). Kalkösung und Karrenbildung. *Zeitschrift für Geomorphologie Supplement Band*, **2**, 4–21.

Bowman, H. N. (1974). *Geology of the Wollongong, Kiama and Robertson 1:50,000 sheets*: Geological Survey of New South Wales.

Braddock, W. A. (1978). Dakota Group rockslides, northern Front Range, Colorado, USA. In *Rockslides and Avalanches, 1, Natural Phenomena*, ed. B. Voight. Amsterdam: Elsevier, pp. 439–480.

Bradley, W. A. (1963). Large-scale exfoliation in massive sandstones of the Colorado Plateau. *Geological Society of America Bulletin*, **75**, 519–528.

Brady, P. V. and Walther, J. V. (1990). Kinetics of quartz dissolution at low temperatures. *Chemical Geology*, **82**, 253–264.

Branagan, D. F. (1983). Tesselated pavements. In *Aspects of Australian Sandstone Landscapes*, eds R. W. Young and G. C. Nanson. Wollongong: Australian and New Zealand Geomorphology Group, pp. 11–20.

Branagan, D. F. (2000). The Hawkesbury Sandstone: its origins and later life. In *Sandstone City, Sydney's Dimension Stone and Other Sandstone Geomaterials*, eds G. H. McNally and B. J. Franklin. Sydney: Geological Society of Australia, pp. 23–38.

Branagan, D. F. and Pedram, H. (1990). The Lapstone structural complex, New South Wales. *Australian Journal of Earth Sciences*, **37**, 23–36.

Braybrooke, J. C. (1990). Some geotechnical phenomena related to high stresses in the Hawkesbury Sandstone. *Proceedings of the 24th Symposium on Advances in the Study of the Sydney Basin*, Newcastle University.

Bremer, H. (1965). Ayers Rock ein Beispiel fur klimatogenetische Morphologie. *Zeitschrift für Geomorphologie*, **9**, 249–284.

Bremer, H. (1980). Landform development in the humid tropics, German geomorphological research. *Zeitscrift für Geomorphologie Supplement Band*, **36**, 162–175.

Bret, M. L. (1962). Les Furnas de Vila-Veilha (Brasil). *Spelunca*, **2**(3), 27–30.

Bretz, J. H. (1923). The channeled scabland of the Columbia Plateau. *Journal of Geology*, **31**, 617–649.

Briceño, H. O. and Schubert, C. (1990). Geomorphology of the Gran Sabana, Guyana Shield, southeastern Venezuela. *Geomorphology*, **3**, 125–141.

Briceño, H. O., Schubert, C. and Paolini, J. (1990). Table-mountain geology and surficial geochemistry: Chimantá Massif Venezuelan Guyana Shield. *Journal of South American Earth Sciences*, **3**, 179–194.

Brighenti, G. (1979). Mechanical behaviour of rocks under fatigue. *Proceedings of the 3rd Congress on Rock Mechanics, Montreux, Vol. 1*, Balkema, Rotterdam, pp. 65–70.

Brinkmann, W. (1986). Particulate and dissolved materials in the Rio Negro–Amazon Basin. In *Sediments and Water Interactions*, ed. G. Sly. New York: Springer, pp. 3–12.

Bromley, R. G. and Mikulas, R. (2002). Sandstone phenomena of East Greenland. *Abstracts of the Conference on Sandstone Landscapes: Diversity, Ecology and Conservation, 14–29 September 2002*, Doubice, Czech Republic.

Brush, L. M. (1961). Drainage basins, channels, and flow characteristics of selected streams in central Pennsylvania. *US Geological Survey Professional Paper*, **282F**, 145–181.

Büdel, J. (1980). Climatic and climatomorphic geomorphology. *Zeitschrift für Geomorphologie Supplement Band*, **36**, 1–8.

Büdel, J. (1982). *Climatic Geomorphology*. Trans. Fischer, L. and Busche, D. Princeton: Princeton University Press.

Burdine, N. T. (1963). Rock failure under dynamic loading conditions. *Society of Petroleum Engineers Journal*, 1–8.

Burford, E. P., Fomina, M. and Gadd, G. M. (2003). Fungal involvement in bioweathering and biotransformation of rocks and minerals. *Mineralogical Magazine*, **67**, 1127–1155.

Burley, S. D. and Kantorowicz, J. D. (1986). Thin section and S.E.M. textural criteria for the recognition of cement-dissolution porosity in sandstones. *Sedimentology*, **33**, 587–604.

Busche, D. (1980). On the origin of the Msak Mallat and Hamadat Manghini Escarpment. In *The Geology of Libya*, eds M. J. Salem and M. T. Busrewil. London: Academic, pp. 837–848.

Busche, D. and Erbe, W. (1987). Silicate karst landforms of the southern Sahara, northeastern Niger and southern Libya. *Zeitschrift für Geomorphologie Supplement Band*, **64**, 55–72.

Busche, D. and Sponholz, B. (1992). Morphological and micromorphological aspects of the sandstone karst of eastern Niger. *Zeitschrift für Geomorphologie Supplement Band*, **85**, 1–18.

Butler, P. R. and Mount, J. F. (1986). Corroded cobbles in southern Death Valley: their relationship to honeycomb weathering and lake shorelines. *Earth Surface Processes and Landforms*, **11**, 377–387.

Campbell, H. and Hutching, G. (2007). *In Search of Ancient New Zealand*. London: Penguin.

Carey, W. S. (1953). The Rheid concept in geotectonics. *Journal of the Geological Society of Australia*, **1**, 67–118.

Carlsson, A. and Olsson, T. (1982). Rock bursting phenomena in a superficial rock mass in southern central Sweden. *Rock Mechanics*, **15**, 99–110.

Carreño, R. and Blanco, F. (2004). Notas sobre la exploración del sistema kárstico de Roraima Sur, Estado Bolívar. *Boletín de la Sociedad Venezolana de Espeleología*, **38**, 45–52.

Carreño, R. and Urbani, F. (2004). Observaciones sobre las espeleotemas del sistema Roraima Sur (in Spanish, English abstract). *Boletín de la Sociedad Venezolana de Espeleología*, **38**, 28–33.

Carter, C. L. and Anderson, R. S. (2006). Fluvial erosion of physically modeled abrasion-dominated slot canyons. *Geomorphology*, **81**, 89–113.

Chalcraft, D. and Pye, K. (1984). Humid tropical weathering of Quartzite in southeastern Venezuela. *Zeitschrift für Geomorphologie*, **28**, 321–332.

Chan, M. A., Seiler, W. M., Ford, R. L. and Yonkee, W. A. (2007). Polygonal cracking and 'Wopmay' weathering patterns on Earth and Mars. *Lunar and Planetary Science*, **38**, 1398.

Chapman, R. M. and Sable, E. G. (1960). Geology of the Utukok–Corwin region, northwestern Alaska. *US Geological Survey Professional Paper*, **303C**.

Cilek, V. (2002). The relief formation of sandstone castelated areas of the Czech Republic. *Abstracts of the Conference on Sandstone Landscapes: Diversity, Ecology and Conservation, 14–20 September 2002*, Doubice in Saxonian-Bohemian Switzerland, Czech Republic.

Cilek, V., Williams, R., Osborne, A., Migon, P. and Mikuláš, R. (2008). The origin and development of sandstone landforms. In *Sandstone Landscapes*, eds H. Hartel, V. Cilek, T. Herben, A. Jackson and R. Williams: Academia, pp. 34–43.

Coates, G. (2002). *The Rise and Fall of the Southern Alps*. Christchurch: Canterbury University Press.

Coates, L. (1989). Micro-organisms and stone: a study of their interaction, with particular emphasis on lichens and sandstone in the Sydney region. Unpublished Master of Geoscience, School of Earth Sciences, Macquarie University.

Colveé, P. (1973). Cueva en Cuarcitas en el Cerro Autana, Territorio Federal Amazonas. *Boletin Sociedad Venezuela Espeleologia*, **4**(1), 5–13.

Conaghan, P. J. (1980). The Hawkesbury Sandstone; gross characteristics and depositional environment. *Geological Survey of New South Wales Bulletin*, **26**, 188–253.

Conca, J. L. and Astor, A. M. (1987). Capillary moisture flow and the origin of cavernous weathering in dolerites of Bull Pass, Antarctica. *Geology*, **15**, 151–154.

Conca, J. L. and Rossman, G. R. (1982). Case hardening of sandstone. *Geology*, **10**, 520–523.

Conca, J. L. and Rossman, G. R. (1985). Core softening in cavernously weathered tonalite. *Journal of Geology*, **93**, 59–73.

Construction Department of Hunan Province (2007). *China Danxia Landform Application for World Natural Heritage – Comprehensive Report, English Version*: Construction Department of Hunan Province, China.

Cooks, J. and Otto, E. (1990). The weathering effects of the lichen *Lecidea Aff. Sarcogynoides (Koerb.)* on Magaliesberg Quartzite. *Earth Surface Processes and Landforms*, **15**, 491–500.

Cooks, J. and Pretorious, J. R. (1987). Weathering basins in the Clarens Formation Sandstone, South Africa. *South African Journal of Geology*, **90**, 147–154.

Coque, R. (1977). *Geomorphologie*. Paris: Colin.

Correa Neto, A. V. (2000). Speleogenesis in quartzite in southeastern Minas Gerais, Brazil. In *Speleogenesis – Evolution of Karst Aquifers*, eds A. Klimchouk, D. C. Ford, A. N. Palmer and W. Dreybrodt. Huntsville: NSS, pp. 452–457.

Correns, C. W. (1941). Über die Löslichkeit von Kieselsäure in schwach sauren und alkalischen Losüngen. *Chemie der Erde*, **13**, 92–96.

Cotton, C. A. (1966). Antarctic scablands. *New Zealand Journal of Geology and Geophysics*, **9**, 130–132.

Coxon, P. (1988). Remnant periglacial features on the summit of Truskmore, Counties Sligo and Leitrim, Ireland. *Zeitschrift für Geomorphologie Supplement Band*, **77**, 81–91.

Crook, K. A. W. (1968). Weathering and roundness of quartz sand grains. *Sedimentology*, **11**, 171–182.

Crozier, M. J., Gage, M., Pettinga, J. R., Selby, M. J. and Wasson, R. J. (1982). The stability of hillslopes. In *Landforms of New Zealand*, eds J. M. Soons and M. J. Selby. Auckland: Longman Paul, pp. 45–56.

Cunningham, D. C. (1988). A rock avalanche in a sandstone landscape, Nattai North, NSW. *Australian Geographer*, **19**, 221–229.

Cunningham, F. F. and Griba, W. (1973). A model of slope development, and its application to the Grand Canyon, Arizona, USA. *Zeitschrift für Geomorphologie*, **17**, 43–77.

Dam, G., Larsen, M. and Sonderholm, M. (1998). Sedimentary response to mantle plumes: implications from Paleocene onshore successions, West and East Greenland. *Geology*, **26**, 207–210.

Dana, J. D. (1850). On the degradation of the rocks of New South Wales and formation of valleys. *American Journal of Science*, **59**, 289–294.

Darwin, C. (1839). *Journal of Researches into the Natural History and Geology of Countries Visited During the Voyage of HMS Beagle Round the World*. London: Ward, Locke & Co. (1890 edition cited).

Davis, S. N. (1964). Silica in streams and ground water. *American Journal of Science*, **262**, 870–891.

Davison, A. P. (1986). An investigation into the relationship between salt weathering debris production and temperature. *Earth Surface Processes and Landforms*, **11**, 335–341.

De Martonne, E. (1927). *Shorter Physical Geography*. London: Christophers.

De Melo, M. S. and Coimbra, A. M. (1996). Ruiniform relief in sandstones: the example of Vila Velha, Carboniferous of the Paraná Basin, Southern Brazil. *Acta Geológica Hispánica*, **31**(4), 25–40.

De Melo, M. S., Godoy, L. C., Meneguzzo, P. M. and Da Silva, D. J. P. (2004). A geologia no plano de manejo do parque estadual de Vila Velha, PR.. *Revista Brasileira de Geociências*, **34** (4), 561–570.

Debossens, G. (2007). The Natural Arches of Tassili National Park, Tassili N'Ajjer, Algeria. Retrieved 24/01/08, from http://www.naturalarches.org/tassili/index.htm.

Delcaillau, B., Herail, G. and Mascle, G. (1987). Evolution geomorphostructurale de fronts de chevauchement actifs: le cas des chevauchements intrasiwaliks du Nepal central. *Zeitschrift für Geomorphologie*, **31**, 339–360.

Denton, C. H., Prentice, M. L., Kellogg, D. E. and Kellogg, T. B. (1984). Late Tertiary history of the Antarctic ice sheet: evidence from the dry Valleys. *Geology*, **12**, 263–267.

Derbyshire, E. (1972). Tors, rock weathering and climate in southern Victoria Land, Antarctica. In *Polar Geomorphology*, eds R. J. Price and D. E. Sugden. London: Institute of British Geographers, pp. 93–106.

Dittes, N. (2002). Field and laboratory testing of St. Peter Sandstone. *Journal of Geotechnical and Geoenvironmental Engineering*, **128**, 372–380.

Dobereiner, L. and de Freitas, M. H. (1986). Geotechnical properties of weak sandstones. *Geotechnique*, **36**, 79–94.

Doerr, S. H. (1997). Hoehlen und andere karstformen in quarziten des Guyanashildes. *El Guacharo*, **40**, 53–74.

Doerr, S. H. (1999). Karst-like landforms and hydrology in quartzites of the Venezuelan Guyana shield: pseudokarst or 'real' karst? *Zeitschrift für Geomorphologie*, **43**, 1–17.

Doerr, S. H. and Wray, R. A. L. (2004). Pseudokarst. In *Encyclopaedia of Geomorphology*, ed. A. S. Goudie. London: Routledge.

Doignon, P. (1973). Une enigme: les pseudosquames polygonales des gres de Fontainebleau. *Bulletin Association Naturale Vallee du Loing*, **49**, 5–9.

Douglas, I. (1969). The efficiency of tropical humid denudation systems. *Transactions of the Institute of British Geographers*, **46**, 1–16.

Douglas, I. (1978). Denudation of silicate socks in the humid tropics. In *Landform Evolution in Australasia*, eds J. Davies and M. A. J. Williams. Canberra: ANU Press, pp. 216–237.

Dove, P. M. and Crerar, D. A. (1990). Kinetics of quartz dissolution in electrolyte solutions using a hydrothermal mixed flow reactor. *Geochimica et Cosmochimica Acta*, **54**, 955–969.

Dove, P. M. and Elston, S. F. (1992). Dissolution kinetics of quartz in sodium chloride solutions: analysis of existing data and a rate model for 25 °C. *Geochimica et Cosmochimica Acta*, **56**, 4147–4156.

Dove, P. M. and Rimstidt, J. D. (1994). Silica–water interactions. In *Silica: Reviews in Mineralogy*, eds P. J. Heaney, C. T. Prewitt and G. V. Gibbs, pp. 259–308.

Dragovich, D. J. (1969). The origin of cavernous surfaces (tafoni) in granitic rocks of southern South Australia. *Zeitschrift für Geomorphologie*, **13**, 163–181.

Dreybrodt, W. (1988). *Processes in Karst Systems*. Berlin: Springer.

Duane, M. J. (2006). Coeval biochemical and biophysical weathering processes on Quaternary sandstone terraces south of Rabat (Teara), northwest Morocco. *Earth Surface Processes and Landforms*, **31**, 1115–1128.

Dury, G. H. (1964). Principles of underfit streams. *US Geological Survey Professional Paper*, **452C**.

Dury, G. H. (1966). Incised valley meanders on the lower Colo River, New South Wales. *Australian Geographical Studies*, **10**, 17–25.

Dusseault, M. B. (1980). Itacolumites: the flexible sandstones. *Quarterly Journal of Engineering Geology London*, **13**, 119–128.

Dutton, C. E. (1882). The physical geology of the Grand Canyon district. *US Geological Survey Annual Report*, **2**, 47–166.

Egboka, B. C. E. and Orajaka, I. P. (1987). The caves of Anambra State: an explanation for their origin. *Journal of the Sydney Speleological Society*, **31**, 3–12.

Fairbridge, R. W. (1988). Cyclical patterns of exposure, weathering and burial of cratonic surfaces, with some examples from North America and Australia. *Geografisker Annaler*, **70A**, 277–283.

Fischer, A. G. (1975). Origin and growth of basins. In *Petroleum and Global Tectonics*, eds A. G. Fischer and S. Judson. Princeton: Princeton University Press, pp. 47–82.

Fisher, D. J., Erdmann, C. E. and Reeside, J. B. (1960). Cretaceous and Tertiary formations of the Book Cliffs, Carbon, Emery and Grand Counties Utah, and Garfield and Mesa Counties Colorado. *US Geological Survey Professional Paper*, **332**.

Fjellanger, J. and Nystuen, J. P. (2007). Diagenesis and weathering of quartzite at the palaeic surface on the Varanger Peninsula, northern Norway. *Norwegian Journal of Geology*, **87**, 133–145.

Fjellanger, J. and Sorbel, L. (2007). Origin of the palaeic landforms and glacial impact on the Varanger Peninsula, northern Norway. *Norwegian Journal of Geology*, **87**, 223–238.

Ford, D. C. (1980). Threshold and limit effects in karst geomorphology. In *Thresholds in geomorphology*, eds D. R. Coates and J. D. Vitek. London: Allen and Unwin.

Ford, D. C. and Lundberg, J. (1987). A review of dissolution rills in limestone and other soluble rocks. *Catena Supplement*, **8**, 119–140.

Franklin, B. J. and Young, J. (2000). The modulus of rupture test and its significance for durability of dimension stone. In *Sandstone City, Sydney's Dimension Stone and Other Sandstone Geomaterials*, eds G. H. McNally and B. J. Franklin. Sydney: Geological Society of Australia, pp. 223–226.

Fränzle, O. (1971). Die Opferkessel im quarztischen Sandstein von Fontainebleau. *Zeitschrift für Geomorphologie*, **15**, 212–235.

Fry, E. J. (1922). Some types of endolithic limestone lichens. *Annals of Botany*, **36**, 541–562.

Fry, E. J. (1924). A suggested explanation of the mechanical action of lithophytic lichens on substrates on rocks. *Annals of Botany*, **38**.

Frye, J. C. and Swineford, A. (1947). Solution features on Cretaceous Sandstone in Central Kansas. *American Journal of Science*, **245**, 366–379.

Furst, M. (1966). Bau und Entstehung der Serir Tibesti. *Zeitschrift für Geomorphologie*, **10**, 387–418.

Gadd, B. (1986). *Handbook of the Canadian Rockies*. Jasper: Corax.

Galán, C., Herrera, F., Carreño, R. and Pérez, M. A. (2004). Roraima Sur System, Venezuela: 10.8 km, world's longest quartzite cave. *Boletín de la Sociedad Venezolana de Espeleología*, **38**, 53–60.

Galloway, R. W. (1976). Geomorphology of the Alligator Rivers area. *CSIRO Australia Land Research Series*, **38**, 52–70.

Gansser, A. (1964). *Geology of the Himalayas*. London: Wiley.

Gansser, A. (1983). *Geology of the Bhutan Himalaya*. Basel: Birkhauser.

George, U. (1989). Venezuela's islands in time. *National Geographic*, **175**, 526–561.

Gerber, E. and Scheidegger, A. E. (1973). Erosional and stress-induced features on steep slopes. *Zeitschrift für Geomorphologie Supplement Band*, **18**, 38–49.

Ghosh, S. K. (1991). Dissolution of silica in nature and its implications. *Bulletin of Canadian Petroleum Geology*, **39**, 212.

Gilbert, G. K. (1896). Niagara Falls and their history. In National Geographic Society, *The Physiography of the United States*. New York: American Book Company, pp. 203–236.

Gilbert, G. K. (1904). Domes and dome structures of the High Sierra. *Geological Society of America Bulletin*, **15**, 29–36.

Godard, A. (1965). *Recherches de geomorphologie en Ecosse du Nord-Ouest*. Paris: Les Belles Lettres.

Goodman, R. E. and Bray, J. W. (1976). Toppling of rock slopes. *Proceedings of the Special Conference on Rock Engineering for Foundations and Slopes, Vol. 2*. New York, American Society of Civil Engineers, pp. 201–234.

Gordon, J. (1981). Ice-scoured topography and its relationship to bedrock structure in parts of northern Scotland and west Greenland. *Geografisker Annaler*, **63A**, 55–65.

Gordon, J. E. (1978). *Structures – Or Why Things Don't Fall Down*. Harmondsworth: Pelican.

Gostin, V. A. and Herbert, C. (1973). Stratigraphy of the Upper Carboniferous and Lower Permian sequence, southern Sydney Basin. *Geological Society of Australia Journal*, **20**, 49–70.

Goudie, A. S. (1974). Further experimental investigation of rock weathering by salt and other mechanical processes. *Zeitschrift für Geomorphologie Supplement Band*, **21**, 1–12.

Goudie, A. S. (1986). Laboratory simulation of the wick effect in salt weathering of rock. *Earth Surface Processes and Landforms*, **11**, 275–285.

Goudie, A. S., Migon, P., Allison, R. J. and Rosser, N. (2002). Sandstone geomorphology of the Al-Quwayra area of south Jordan. *Zeitschrift für Geomorphologie*, **46**, 365–390.

Graf, W. L., Hereford, R., Laity, J. and Young, R. A. (1987). Colorado Plateau. In *Geomorphic Systems of North America*, ed. W. L. Graf. Washington: Geological Society of America, pp. 259–302.

Gregory, H. E. (1917). Geology of the Navajo Country. *US Geological Survey Professional Paper*, **93**.

Gregory, H. E. (1938). The San Juan Country. *US Geological Survey Professional Paper*, **188**.

Gregory, H. E. (1950). Geology and geography of the Zion Park region Utah and Arizona. *US Geological Survey Professional Paper*, **220**.

Gregory, H. E. and Moore, R. C. (1931). The Kaiparowits region. *US Geological Survey Professional Paper*, **164**.

Griffith, A. A. (1921). The phenomena of rupture and flow in solids. *Philosophical Transactions of the Royal Society*, **A221**, 163–198.

Griggs, D. T. (1936). The factor of fatigue in rock exfoliation. *Journal of Geology*, **44**, 783–796.

Grimes, K. G. (2007). *Whalemouth Cave, WA, an example of tropical sandstone karst.* Hamilton: CD Resource, Regolith Mapping.

Grunert, J. and Busche, D. (1980). Large-scale fossil landslides at the Msak Malat and Hamadat Manghini Escarpment. In *The Geology of Libya*, eds M. J. Salem and M. T. Busrewil. London: Academic, pp. 849–860.

Hack, J. T. (1957). Studies of longitudinal stream profiles in Virginia and Maryland. *US Geological Survey Professional Paper*, **294B**, 45–97.

Hack, J. T. (1960). Interpretation of erosional topography in humid temperate regions. *American Journal of Science*, **258A**, 80–97.

Hack, J. T. (1965). Geomorphology of the Shenandoah valley, and origin of the residual ore deposits. *US Geological Survey Professional Paper*, **484**.

Hagedorn, H. (1971). Untersuchungen uber Relieftypen arider Raume an Beispelen aus dem Tibesti-Gebirge und seiner Umgebung. *Zeitschrift für Geomorphologie Supplement Band*, **11**, 1–251.

Haimson, B. C. (1974). Mechanical behaviour of rock under cyclical loading. *Proceedings of the 3rd Congress of the International Association for Rock Mechanics, Vol. 2*, Denver, National Academy of Science, Washington.

Hale, M. E. (1987). Epilithic lichens in the Beacon Sandstone formation, Victoria Land, Antarctica. *Lichenologist*, **19**, 2269–2287.

Hall, A. M. (1986). Deep weathering patterns in north–east Scotland and their geomorphological significance. *Zeitschrift für Geomorphologie*, **30**, 407–422.

Haller, J. (1971). *Geology of the East Greenland Caledonides*. London: Wiley.

Halliday, W. R. (2003). Caves and karst of northeast Africa. *International Journal of Speleology*, **32**, 19–32.

Hambrey, M. J., Glasser, N. F., McKelvey, B. C., Sugden, D. E. and Fink, D. (2007). Cenozoic landscape evolution of an East Antarctic oasis (Radok Lake area, northern Prince Charles Mountains), and its implications for the glacial and climatic history of Antarctica. *Quaternary Science Reviews*, **26**, 598–626.

Hancock, G. and Kirwan, M. (2007). Summit erosion rates deduced from [10]Be: implications for relief production in the central Appalachians. *Geology*, **35**, 89–92.

Hardy, H. R. and Chugh, Y. P. (1971). Failure of geological material under low-cycle fatigue. *Proceedings of the 6th Canadian Rock Mechanics Symposium, Montreal*, Balkema, Rotterdam, pp. 33–47.

Harshbarger, C. A., Repenning, C. A. and Irwin, J. H. (1957). Stratigraphy of the uppermost Triassic rocks and Jurassic rocks of the Navajo country. *US Geological Survey Professional Paper*, **291**.

Haynes, V. M. (1968). The influence of glacial erosion and rock structure on corries in Scotland. *Geografisker Annaler*, **50A**, 221–234.

Heald, M. T. and Anderegg, R. C. (1960). Differential cementation in the Tuscarora Sandstone. *Journal of Sedimentary Research*, **30**, 568–577.

Hedges, J. (1969). Opferkessel. *Zeitschrift für Geomorphologie*, **13**, 22–55.

Henderson, M. E. K. and Duff, R. B. (1963). The release of metallic and silicate ions from minerals, rocks, and soils by fungal activity. *Journal of Soil Science*, **14**, 236–246.

Henderson, P. (1982). *Inorganic Geochemistry*. Oxford: Pergamon.

Herget, G. (1988). *Stresses in Rock*. Rotterdam: Balkema.

Hettner, A. (1887). Gebirgsbau und oberflachengestaltung der Sachischen Schweiz. *Forschung zur deutsches Landes-und Volkskunde*, **2**, 245–355.

Hettner, A. (1903). Die Felsbildungen der sachsischen Schweiz. *Geographische Zeitschrift*, **9**, 608–626.

Hettner, A. (1928). *The Surface Features of the Land: Problems and Methods of Geomorphology*. Trans. Tilley, P. (1972), New York: Hafner.

Heyman, J. (1982). *The Masonry Arch*. Chichester: Ellis Harwood.

Hill, C. A. and Forti, P. (1986). *Cave Minerals of the World*. Huntsville, USA: National Speleological Society.

Hoatson, D., Blake, D., Mory, A., Tyler, I., Pittavino, M., Allen, B., Kamprad, J. and Oswald-Jacobs, I. (1997). *Bungle Bungle Range, Purnululu National Park, Western Australia: A Guide to the Rocks, Landforms, Plants, Animals and Human Impact*. Canberra: Australian Geological Survey Organisation.

Hocking, R. M. (1980). The Tumblagooda Sandstone, Western Australia. *Report of the Geological Survey of Western Australia*, pp. 53–61.

Hoek, E. (1968). Brittle failure of rock. In *Rock Mechanics in Engineering Practice*, eds K. G. Stagg and O. C. Zienkiewicz. London: Wiley, pp. 99–124.

Hoek, E. (1983). Strength of jointed rock masses. *Geotechnique*, **33**, 187–223.

Hoek, E. and Bray, J. (1974). *Rock Slope Engineering*. London: Institute of Mining and Metallurgy.

Hoek, E. and Brown, E. T. (1980). *Underground Excavations in Rocks*. London: Institute of Mining and Metallurgy.

Hoek, E. and Brown, E. T. (1997). Practical estimates of rock mass strength. *International Journal of Rock Mechanics and Mining Science*, **34**, 1165–1186.

Hoffland, E., Kuyper, T. W., Wallander, H., Plassard, C., Gorbushina, A. A., Haselwandter, K., Holmström, S., Landeweert, R., Lundström, U. S., Rosling, A., Sen, R., Smits, M. M., van Hees, P. A. W. and van Breemen, N. (2004). The role of fungi in weathering. *Frontiers in Ecology and the Environment*, **2**(5), 258–264.

Hoffman, L. and Darienko, T. (2005). About algal growth on the sandstone outcrops of Luxembourg. *Sandstone Landscapes in Europe – Past, Present and Future. Proceedings of the 2nd Conference on Sandstone Landscapes, 25–28.05.2005*, Vianden (Luxembourg), Ferrantia 44, 249 p.

Holland, W. M. (1977). Slot valleys. *Australian Geographer*, **13**, 338–339.

Holtedahl, O. (1960). Geology of Norway. *Norge Geologiske Undersokelse*, **208**, Oslo.

Horton, B. K. and De Celles, P. G. (2001). Modern and ancient fluvial migrations in the foreland basin system of the central Andes, southern Bolivia: implications for drainage network evolution in fold–thrust belts. *Basin Research*, **13**, 43–63.

Howard, A. D. and Kochel, R. C. (1988). Introduction to cuesta landforms and sapping processes on the Colorado Plateau. In *Sapping Features of the Colorado Plateau*, eds A. D. Howard, R. C. Kochel and H. E. Holt. Washington: NASA, pp. 6–56.

Huang, W. H. and Keller, W. D. (1970). Dissolution of rock-forming silicate minerals in organic acids: simulated first stage weathering of fresh mineral surfaces. *The American Mineralogist*, **55**, 2076–2094.

Hubert, H. (1920). Grottes et cavernes de l'Afrique Occidentale. *Bulletin de Comité d'Etudes Historiques et Scientifiques de l' Afrique Occidentale Francaise*, 43–51.

Humphreys, P. and Mitchell, P. (1983). A preliminary assessment of the role of bioturbation and rainwash on sandstone hillslopes in the Sydney Basin. In *Aspects of Australian Sandstone Landscapes*, eds R. W. Young and G. C. Nanson. Wollongong: Australian and New Zealand Geomorphology Group, pp. 66–80.

Huntoon, P. W. (1989). Gravity tectonics, Grand Canyon, Arizona. In *Geology of Grand Canyon, Northern Arizona*, eds D. P. Elston, G. H. Billingsley and R. A. Young. Washington: American Geophysical Union, pp. 219–233.

Hurst, A. and Bjorkum, P. A. (1986). Discussion, thin section and SEM textural criteria for the recognition of cement-dissolution porosity in sandstones. *Sedimentology*, **33**, 605–614.

Iller, R. K. (1979). *Chemistry of Silica*. New York: Wiley.

Ives, R. L. (1947). Reconnaissance of the Zion hinterland. *Geographical Review*, **37**, 618–638.

Jackson, K. S., Jonasson, I. R. and Skippen, G. B. (1978). The nature of metals–sediment–water interactions in freshwater bodies, with emphasis on the role of organic matter. *Earth Science Reviews*, **14**, 97–146.

Jaeger, J. C. (1971). Friction of rocks and stability of rock slopes. *Geotechnique*, **21**, 97–134.

Jaeger, J. C. and Cook, N. G. (1969). *Fundamentals of Rock Mechanics*. London: Chapman and Hall.

Jaggar, F. (1978a). *Rock mechanics investigation of the structural stability in the Bulli seam at West Cliff Colliery*. Wollongong: Australian Coal Industry Research Laboratory.

Jaggar, F. (1978b). *Rock mechanics investigation of structural stability in the Greta Seam at Pelton Colliery*. Wollongong: Australian Coal Industry Research Laboratory.

Jamison, W. R. and Stearns, D. W. (1982). Tectonic deformation on Wingate Sandstone, Colorado National Monument. *Association of American Petroleum Geologists*, **66**, 2584–2608.

Jennings, J. N. (1967). Further remarks on the Big Hole, near Braidwood, New South Wales. *Helictite*, **6**, 3–9.

Jennings, J. N. (1979). Arnhem Land city that never was. *Geographical Magazine*, **60**, 822–827.

Jennings, J. N. (1983). Sandstone pseudokarst or karst? In *Aspects of Australian Sandstone Landscapes*, eds R. W. Young and G. C. Nanson. Wollongong: Australian and New Zealand Geomorphology Group, pp. 21–30.

Jennings, J. N. (1985). *Karst Geomorphology*. Oxford: Basil Blackwell.

Jerzykiewicz, T. and Wojewoda, J. (1986). The Radkow and Szczeliniec sandstones; an example of giant foresets on a tectonically controlled shelf of the Bohemian Cretaceous basin. In *Shelf Sands and Sandstones*, eds R. J. Knight and J. McLean. Calgary: Canadian Society of Petroleum Geologists, pp. 1–15.

Johnson, D. (2004). *The Geology of Australia*. Cambridge: Cambridge University Press.

Johnson, M. R. (1981). The erosional factor in the emplacement of the Keystone thrust sheet (South East Nevada) across a land surface. *Geological Magazine*, **118**, 501–507.

Johnson, R. H. and Vaughan, R. D. (1983). The Alport Castles, Derbyshire: a south Pennine slope and its geomorphic history. *East Midlands Geographer*, **8**, 79–88.

Johnson, W. (1960). Geological investigations at Warragamba Dam to the end of 1955. *Journal of the Australian Institute of Engineers*, **32**, 85–97.

Johnsson, M. J., Stallard, R. F. and Meade, R. H. (1988). First-cycle quartz arenites in the Orinoco River Basin, Venezuela and Colombia. *Journal of Geology*, **96**, 263–277.

Jones, B. G. and Rust, B. R. (1983). Massive sandstone facies in the Hawkesbury sandstone, a Triassic fluvial deposit near Sydney, Australia. *Journal of Sedimentary Petrology*, **53**, 1249–1259.

Jones, J. B. and Segnit, E. R. (1971). The nature of opal: 1. Nomenclature and constituent phases. *Journal of the Geological Society of Australia*, **18**, 57–68.

Journaux, A. and Coutard, J. P. (1974). Experiences de thermoclastie sur roches siliceuses. *Centre de geomorphologie de Caen C.N.R.S. Bulletin*, **18**, 7–20.

Joyce, E. B. (1974). The sandstone caves of Mt Moffatt Station, southern Queensland, Australia – speleogenesis, cave deposits and aboriginal occupation. *5th International Kongress für Speleology, Bd2 1969*, Stuttgart.

Jukes, J. B. (1850). *A Sketch of the Physical Structure of Australia, So Far as it is at Present Known*. London: Boone.

Julien, A. A. (1879). On the geological action of the humus acids. *Proceedings of the American Association for the Advancement of Science*, **28**, 311–410.

Kastner, M. (1981). Authigenic silicates in deep sea sediments: Formation and diagenesis. In *The Oceanic Lithosphere*, ed. C. Emiliani. New York: John Wiley & Sons, pp. 915–980.

Kawamoto, T. and Fujita, M. (1968). State of stress and deformation of natural slopes, Tsuchi to Kiso. *Japanese Society for Soil Mechanics and Foundation Engineering*, **16**, 37–46 (in Japanese, English summary).

Keefer, D. K. (1984). Landslides caused by earthquakes. *Geological Society of America Bulletin*, **95**, 406–421.

Kelly, J. L., Fu, B., Kita, N. T. and Valley, J. W. (2007). Optically continuous silcrete quartz cements of the St. Peter Sandstone: high precision oxygen isotope analysis by ion microprobe. *Geochimica et Cosmochimica Acta*, **71**, 3812–3832.

Kelly, V. C. (1955). Monoclines of the Colorado Plateau. *Geological Society of America Bulletin*, **66**, 789–804.

King, L. C. (1959). Denudational and tectonic relief in southeastern Australia. *Transactions of the Geological Society of South Africa*, **62**, 113–138.

King, L. C. (1962). *The Morphology of the Earth*. Edinburgh: Oliver and Boyd.

Kirkbride, M. and Matthews, D. (1997). The role of fluvial and glacial erosion in landscape evolution: the Ben Ohau Range, New Zealand. *Earth Surface Processes and Landforms*, **22**, 317–327.

Koons, E. D. (1955). Cliff retreat in the southwestern United States. *American Journal of Science*, **253**, 44–52.

Krauskopf, K. B. (1956). Dissolution and precipitation of silica at low temperatures. *Geochimica et Cosmochimica Acta*, **10**, 1–26.

Krynine, P. D. (1941). Differentiation of sediments during the life history of a landmass (abstract). *Geological Society of America Bulletin*, **52**, 1915.

Kumar, R. and Kumar, A. V. (1999). *Biodeterioration of Stone in Tropical Environments: An Overview*: J. Paul Getty Trust.

Kurtz, H. D. and Netoff, D. I. (2001). Stabilization of friable sandstone surfaces in a desiccating, wind-abrading environment of southcentral Utah by rock surface microorganisms. *Journal of Arid Environments*, **48**, 89–100.

Laffer, L. (1973). La Cueva de Conjero. *El Guacharo*, **16**, 31–34.

Laity, J. (1987). The Colorado Plateau in planetary geology studies. In *The Geomorphic Systems of North America*, ed. W. L. Graf. Boulder: Geological Society of America, pp. 288–297.

Laity, J. (1988). The role of groundwater sapping in valley evolution on the Colorado Plateau. In *Sapping Features of the Colorado Plateau*, eds A. D. Howard, R. C. Kochel and H. E. Holt. Washington: NASA, pp. 63–70.

Laity, J. and Malin, M. C. (1985). Sapping processes and the development of theater-headed valley networks on the Colorado Plateau. *Geological Society of America Bulletin*, **96**, 203–217.

Lajtai, E. Z. (1971). A theoretical and experimental evaluation of the Griffith theory of brittle fracture. *Tectonophysics*, **11**, 129–156.

Lama, R. D. and Vukuturi, V. S. (1978). *Handbook on Mechanical Properties of Rocks, Vol. 14* Aedermannsdorf: Tran Tech.

Lassak, E. V. (1970). A note on some non-calcareous stalactites from sandstones of the Sydney Basin, N.S.W. *Journal and Proceedings of the Royal Society of New South Wales*, **103**, 11–14.

Latocha, A. and Synowiec, G. (2002). Comparison of the sandstone landscapes of the Stolowe and Bzstrzyckie Mountains, Sudetes, SW Poland. *Abstracts of the Conference on Sandstone Landscapes: Diversity, Ecology and Conservation, 14–20 September 2002*, Doubice in Saxonian-Bohemian Switzerland, Czech Republic.

Latocha, A. and Synowiec, G. (2008). Comparison of the sandstone landscapes of the Stolowe and Bystrzyckie Gory Mountains, Sudetes (SW Poland). In *Sandstone Landscapes*, eds H. Hartel, V. Cilek, T. Herben, A. Jackson and R. Williams. New York: Academia, pp. 56–60.

Le Roux, J. P. and Jones, B. G. (1994). Lithostratigraphy and depositional environment of the Permian Nowra Sandstone in the southwestern Sydney Basin, Australia. *Australian Journal of Earth Sciences*, **41**, 191–203.

Leaman, D. E. (1990). The Sydney Basin: composition of the basement. *Australian Journal of Earth Sciences*, **37**, 107–108.

Lewis, A. R., Marchant, D. R., Kowalski, D. E., Baldwin, S. L. and Webb, L. E. (2006). The age and origin of the Labyrinth, western Dry Valleys, Antarctica: evidence for extensive middle Miocene subglacial floods and freshwater discharge to the Southern Ocean. *Geology*, **34**, 513–516.

Li, Y. H. (1975). Denudation of Taiwan island since the Pliocene Epoch. *Geology*, **4**, 105–107.

Livingstone, D. (1963). Chemical composition of lakes and rivers. *US Geological Survey Professional Paper*, **400G**.

Löffler, E. (1977). *Geomorphology of Papua New Guinea*. Canberra: ANU Press.

Longwell, C. R. (1949). Structure of the northern Muddy Mountain area, Nevada. *Geological Society of America Bulletin*, **60**, 923–968.

Louis, H. (1964). Über Rumpflachen und Talbildung in den wechselfeuchten Tropen, besonders nach Studien in Tanganjika. *Zeitschrift für Geomorphologie*, **8**, 43–70.

Louis, H. and Fischer, K. (1979). *Allgemeine Geomorphologie*. Berlin: de Gruyter.

Lozano, M. V., Fabregat, C., López, S. and González, J. M. (2008) Albarracin Rodeno Landscape (Spain). In *Sandstone Landscapes*, eds H. Hastel, V. Click, T. Herben, A. Jackson and R. Williams. New York: Academia, pp. 368–371.

Lyakhnitsky, Y. S. and Khlebalin, I. Y. (2006). Pseudokarst and non-karst caves – A discussion and examples from Russia. *Abstracts of the 9th International Symposium on Pseudokarst*, 24–26 May 2006, Bartkowa, Poland, Institute of Nature Conservation, Polish Academy of Sciences, Cracow.

Mabbutt, J. A. (1962). Geomorphology of the Alice Springs area. *CSIRO Australia Land Research Series*, **6**, 163–184.

Mackel, R. (1974). Dambos – a study of morphodynamic activity on the plateau regions of Zambia. *Catena*, **1**, 327–336.

Mackel, R. (1985). Dambos and related landforms in Africa. *Zeitschrift für Geomorphologie Supplement Band*, **52**, 1–23.

Mainguet, M. (1966). Le Borkou, aspects d'un modele eolien. *Annales de Geographie*, **421**, 296–322.

Mainguet, M. (1970). Un etonnant paysage: le cannelures greseuses du Bembeche (N. du Tchad); essai d'explication geomorphologique. *Annales de Geographie*, **79**, 58–66.

Mainguet, M. (1972). *Les Modelé des Grès*. Paris: Institut Géographique National.

Margielewski, W. and Urban, J. (2003). Crevice-type caves as initial forms of rock landslide development in the Flysch Carpathians. *Geomorphology*, **54**, 325–338.

Marker, M. E. (1976). Note on some South African pseudokarst. *Boletin Sociedad Venezuela Espeleologia*, **7**(13), 5–12.

Marker, M. E. and Swart, P. G. (1995). Pseudokarst in the Western Cape, South Africa: its palaeoenvironmental significance. *Cave and Karst Science, Transactions of the British Cave Research Association*, **22**, 31–38.

Marshall, W. L. and Warakomski, J. M. (1980). Amorphous silica solubilities – II. Effect of aqueous salt solutions at 25 °C. *Geochimica et Cosmochimica Acta*, **44**, 915–924.

Martin, H. A. (1978). Evolution of the Australian flora and vegetation through the Tertiary; evidence from pollen. *Alcheringa*, **2**, 181–202.

Martini, J. E. J. (1979). Karst in Black Reef Quartzite near Kaapsehoop, Eastern Transvaal. *Annals of the South African Geological Survey*, **13**, 115–128.

Martini, J. E. J. (1982). Karst in Black Reef and Wollenberg Group Quartzite of the Eastern Transvaal Escarpment. *Boletin Sociedad Venezuela Espeleologia*, **10**, 99–114.

Martini, J. E. J. (2000a). Dissolution of quartz and silicate minerals. In *Speleogenesis – evolution of Karst Aquifers*, eds A. Klimchouk, D. C. Ford, A. N. Palmer and W. Dreybrodt. Huntsville: NSS.

Martini, J. E. J. (2000b). Quartzite caves in Southern Africa. In *Speleogenesis – Evolution of Karst Aquifers*, eds A. B. Klimchouk, D. C. Ford, A. N. Palmer and W. Dreybrodt. Huntsville: NSS.

Matmon, A., Shaked, Y., Porat, N., Enzel, Y., Finkel, R., Lifton, N., Boaretto, E. and Agnon, A. (2005). Landscape development in an hyperarid sandstone environment along the margins of the Dead Sea fault: implications from dated rock falls. *Earth and Planetary Science Letters*, **240**, 803–817.

Mbanugoh, E. and Egboka, B. C. E. (1988). Hydrogeotectonic origin of the Ogbunike Cave, Anambra State, Nigeria. *Journal of the Sydney Speleological Society*, **32**(4), 65–75.

McFarlane, M. J. and Twidale, C. R. (1987). Karstic features associated with tropical weathering profiles. *Zeitschrift für Geomorphology Supplement Band*, **64**, 73–95.

McGill, G. E. and Stromquist, A. W. (1979). The grabens of Canyonlands National Park, Utah; geometry, mechanics and kinematics. *Journal of Geophysical Research*, **84**, 4547–4563.

McGreevy, J. P. (1982). 'Frost and salt' weathering: further experimental results. *Earth Surface Processes and Landforms*, **7**, 475–488.

McGreevy, J. P. (1985). A preliminary scanning electron microscope study of honeycomb weathering of sandstone in a coastal environment. *Earth Surface Processes and Landforms*, **10**, 509–518.

McGreevy, J. P. and Smith, B. J. (1984). The possible role of clay minerals in salt weathering. *Catena*, **11**, 169–175.

McKee, E. D. and Resser, C. E. (1945). *Cambrian History of the Grand Canyon Region*. Carnegie Institute, Washington, Publication 563.

McNally, G. H. (1981). Valley bulging, Mangrove Creek dam near Gosford, NSW. *Geological Survey of New South Wales Quarterly Notes*, **42**, 4–11.

McNally, G. H. (1993). Effects of weathering on the engineering properties of the Terrigal Formation. In *Collected Case Studies in Engineering Geology, Hydrogeology and Environmental Geology*, eds G. McNally, M. Knight and R. Smith. Springwood: Butterfly Books, pp. 92–102.

McNally, G. H. and McQueen, L. (2000). The geomechanical properties of sandstones and what they mean. In *Sandstone City, Sydney's Dimension Stone and Other Sandstone Geomaterials*, eds G. H. McNally and B. J. Franklin. Sydney: Geological Society of Australia, pp. 178–196.

McQueen, L. B. (2000). Stress relief effects in sandstones in Sydney underground and deep excavations. In *Sandstone City, Sydney's Dimension Stone and Other Sandstone Geomaterials*, eds G. H. McNally and B. J. Franklin. Sydney: Geological Society of Australia, pp. 309–329.

McSaveney, M. J. (1978). Sherman Glacier rock avalanche. In *Rock Slides and Avalanches, 1, Natural Phenomena*, ed. B. Voight. Amsterdam: Elsevier, pp. 197–258.

Mecchia, M. and Piccini, L. (1999). Hydrogeology and SiO_2 geochemistry of the Aonda cave system, Auyán-Tepui, Boívar, Venezuela. *Boletín de la Sociedad Venezolana de Espeleología*, **33**, 1–18.

Mercier, D. (2002). La dynamique paraglaciere des versants du Svalbard. *Zeitschrift für Geomorphologie*, **46**, 203–222.

Mertík, J. and Adamovic, J. (2005). Some significant geomorphic features of the Klokoci Cuesta, Czech Republic. *Sandstone Landscapes in Europe – Past, Present and Future. Proceedings of the 2nd Conference on Sandstone Landscapes, 25–28.05.2005*, Vianden (Luxembourg), Ferrantia 44, 249 p.

Meybeck, M. (1987). Global chemical weathering of surficial rocks estimated from river dissolved loads. *American Journal of Science*, **287**, 401–428.

Michael, E. D. (1965). Origin of the lakes in the Chuska Mountains, northwestern New Mexico; discussion. *Geological Society of America Bulletin*, **76**, 267–268.

Migon, P. (in prep.). Geomorphological Landscapes of the World.

Migon, P., Tulaczyk, S. and Rozpendowski, G. (2002). Sandstone landscapes in the NW part of the Intrasudetic Depression, Sudetes Mountains. *Abstracts of the Conference on Sandstone Landscapes: Diversity, Ecology and Conservation, 14–20 September 2002*, Doubice in Saxonian-Bohemian Switzerland, Czech Republic.

Mikulás, R. (2008a). Microforms of the sandstone relief. In *Sandstone Landscapes*, eds H. Hartel, V. Cilek, T. Herben, A. Jackson and R. Williams. New York: Academia, pp. 66–75.

Mikulás, R. (2008b). The concept of porokarst: sandstone sculpturing across climatic zones and lithofacies. In *Sandstone Landscapes*, eds H. Hartel, V. Cilek, T. Herben, A. Jackson and R. Williams. New York: Academia, pp. 79–82.

Miller, H. (1889). *The Old Red Sandstone*. Edinburgh: Nimmo, Hay and Mitchell.

Mills, H. H. (1981). Boulder deposits and the retreat of mountain slopes, or 'gully gravure' revisited. *Journal of Geology*, **189**, 649–660.

Mocchiutti, A. and Maddaleni, P. (2005). Chemical, geomechanical and geomorphological aspects of karst in sandstone and marl of flysch formations in North East Italy. *Acta Carstologica*, **34**(2), 349–368.

Moon, B. P. and Selby, M. J. (1983). Rock mass strength and scarp forms in southern Africa. *Geografiska Annaler*, **65A**, 135–145.

Morris, R. C. and Fletcher, A. B. (1987). Increased solubility of quartz following ferrous–ferric iron reactions. *Nature*, **330**(6148), 558–561.

Morse, J. W. and Casey, W. H. (1988). Ostwald processes and mineral paragenesis in sediments. *American Journal of Science*, **288**, 537–560.

Mottershead, D. N., Gorbushina, A. A., Lucas, G. R. and Wright, J. (2003). The influence of marine salts, aspect and microbes in the weathering of sandstone in two historic structures. *Building and Environment*, **38**, 1193–1204.

Mullan, G. J. (1989). Caves of the Fell Sandstone of Northumberland. *Proceedings of the University of Bristol Speleological Society*, **18**, 430–437.

Mullis, A. M. (1991). The role of silica precipitation kinetics in determining the rate of Quartz pressure solution. *Journal of Geophysical Research*, **96**, 10,007– 10,013.

Munro-Perry, P. M. (1990). Slope development in the Klerkspruit valley, Orange Free State, South Africa. *Zeitschrift für Geomorphologie*, **34**, 409–421.

Mustoe, G. E. (1983). Cavernous weathering in the Capitol Reef desert, Utah. *Earth Surface Processes and Landforms*, **8**, 517–526.

Nesbitt, H. W., Fedo, C. M. and Young, G. M. (1997). Quartz and feldspar stability, steady and non-steady state weathering, and petrogenesis of siliciclastic sands and muds. *Journal of Geology*, **105**, 173–191.

Nesje, A. and Whillans, I. M. (1994). Erosion of Sognefjord, Norway. *Geomorphology*, **9**, 33–45.

Netoff, D. I. (1971). Polygonal jointing in sandstone near Boulder, Colorado. *Rocky Mtns. Association of Geologists*, **18**, 17–24.

Netoff, D. I. and Shroba, R. S. (1993). Giant weathering pits in the Entrada Sandstone, Southeastern Utah: preliminary findings. *Geological Society of America Abstract*, **25598**.

Netoff, D. I. and Shroba, R. S. (2001). Conical sandstone landforms cored with clastic pipes in Glen Canyon National Recreation Area, southeastern Utah. *Geomorphology*, **39**, 99–110.

Nicholas, R. M. and Dixon, J. C. (1986). Sandstone scarp form and retreat in the Land of Standing Rocks, Canyonlands National Park, Utah. *Zeitschrift für Geomorphologie*, **30**, 167–187.

Nott, J. F. (2006). *Extreme Events, A Physical Reconstruction and Risk Assessment*. Cambridge: Cambridge University Press.

Nott, J. F. and Owen, J. A. K. (1992). An Oligocene palynoflora from the middle Shoalhaven catchment N.S.W. and the Tertiary evolution of flora and climate in the southeast Australian highlands. *Palaeogeography, Palaeoclimatology, Palaeoecology*, **95**, 135–151.

Nott, J. F. and Ryan, P. (1996). Large-scale, subcircular depressions across western Arnhem Land interpreted as exhumed giant lunate current ripples. *Australian Journal of Earth Sciences*, **43**.

Nott, J. F., Young, R. W. and McDougall, I. (1996). Wearing down, wearing back and gorge extension in the long-term denudation of a highland mass: quantitative evidence from the Shoalhaven catchment, southeast Australia. *Journal of Geology*, **104**, 224–232.

Oberlander, T. M. (1977). Origin of segmented cliffs in massive sandstones of southeastern Utah. In *Geomorphology of Arid Regions*, ed. D. O. Doehring. Boston: Allen and Unwin, pp. 79–114.

Oberlander, T. M. (1989). Slope and pediment systems. In *Arid Zone Geomorphology*, ed. D. S. Thomas. London: Belhaven, pp. 56–84.

Okamoto, G., Okura, T. and Goto, K. (1957). Properties of silica in water. *Geochimica et Cosmochimica Acta*, **12**, 123–132.

Ollier, C. D. (1969). *Weathering*. Edinburgh: Oliver and Boyd.

Ollier, C. D. and Tuddenham, W. G. (1961). Inselbergs of central Australia. *Zeitschrift für Geomorphologie*, **5**, 257–276.

Ori, G. G. and Roveri, M. (1987). Geometries of Gilbert-type deltas and large channels in the Meteora Conglomerate, Meso-Hellenic basin (Oligo-Miocene), central Greece. *Sedimentology*, **34**, 845–859.

Orwin, J. F. (1998). The application and implications of rock weathering–rind dating to a large rock avalanche, Craigieburn Range, Canterbury, New Zealand. *New Zealand Journal of Geology and Geophysics*, **41**, 219–223.

Osborn, G. (1985). Evolution of the late Cenozoic inselberg landscape of southwestern Jordan. *Palaeogeography, Palaeoclimatology, Palaeoecology*, **149**, 1–23.

Palmer, J. and Radley, J. (1961). Gritstone tors of the English Pennines. *Zeitschrift für Geomorphologie*, **5**, 37–52.

Papida, S., Murphy, M. and May, E. (2000). Enhancement of physical weathering of building stones by microbial populations. *International Biodeterioration and Biodegradation* **46**, 305–317.

Paradise, T. R. (1997). Disparate sandstone weathering beneath lichens, Red Mountain, Arizona. *Geografisker Annaler*, **79A**, 177–184.

Paradise, T. R. (2003). Sandstone weathering and aspect in Petra, Jordan. *Zeitschrift für Geomorphologie* **46**, 1–17.

Paton, T. R., Humphreys, G. S. and Mitchell, P. B. (1995). *Soils – A New Global View*. London: UCL Press.

Patton, W. W. and Tailleur, I. L. (1964). Geology of the Killik–Itkillik region, Alaska. *US Geological Survey Professional Paper*, **303G**.

Pavey, A. J. (1972). Hilltop natural tunnel. *SPAR, Newsletter of the University of New South Wales Speleological Society*, **17**(July), 17–18.

Pells, P. J. (1977). Measurement of engineering properties of the Hawkesbury Sandstone. *Australian Geomechanics Journal*, **1**, 10–20.

Pells, P. J. (1985). Engineering properties of the Hawkesbury sandstone. In *Engineering Geology of the Sydney Region*, ed. P. J. Pells. Rotterdam: Balkema, pp. 179–197.

Penck, W. (1924). *Die Morphologische Analyse (Morphological Analysis of Landforms)*. Trans. Czech, H. and Boswell, K. New York: Hafner.

Peng, H. (2000). *Danxia geomorphology of China and its progress in research work*. Zhongshan University.

Peng, H. (2007). *The Red Stone Park of China, Danxiashan*. Beijing: Geological Publishing House.

Pettijohn, F. J., Potter, P. E. and Siever, R. (1972). *Sand and Sandstone*. Berlin: Springer.

Pettinga, J. R. (1987). Ponui landslide: a deep-seated wedge failure in Tertiary weak-rock flysch, southern Hawkes Bay, New Zealand. *New Zealand Journal of Geology and Geophysics*, **30**, 415–430.

Philbrick, S. (1970). Horizontal configuration and the rate of erosion of Niagara Falls. *Geological Society of America Bulletin*, **81**, 3723–3732.

Phillips, W. M., Comins, D. G., Gupta, S. and Kubic, P. W. (2004). Rates of knickpoint migration and bedrock erosion from cosmogenic [10]Be in a landscape of active normal faulting. *Geophysical Research Abstracts*, **6**, 6582.

Piccini, L. (1995). Karst in siliceous rocks: karst landforms and caves in the Auyán-tepui massif (Est. Bolivar, Venezuela). *International Journal of Speleology*, **24**, 2–13.

Pickard, J. and Jacobs, S. W. (1983). Vegetation patterns on the Sassafras Plateau. In *Aspects of Australian Sandstone Landscapes*, eds R. W. Young and G. C. Nanson. Wollongong: Australian and New Zealand Geomorphology Group, pp. 54–65.

Pitteau, D. R. and Jennings, J. E. (1970). The effects of plan geometry on the stability of natural slopes in rock in the Kimberley area of South Africa. *Proceedings of the 2nd Congress of the International Society of Rock Mechanics*, Privredni pregled, Belgrade, pp. 289–295.

Porter, W. P. (1979). Opaline outgrowths on Lee Formation Sandstones in Wise County, Virginia. *Rocks and Minerals*, **54**, 97–100.

Posamentier, H. W. and Allen, G. P. (1993). Variability of the sequence stratigraphic model: effects of local basin factors. *Sedimentary Geology*, **86**, 91–109.

Pouyllau, M. and Seurin, M. (1985). Pseudo-karst dans des roches gréso-quartzitiques de la formation Roraima. *Karstologia*, **5**, 45–52.

Powell, J. W. (1895). *Canyons of the Colorado*. Meadville: Flood and Vincent.

Priest, S. D. and Selvakumar, S. (1982). *The Failure Characteristics of Selected British Rocks*. London: Report for the Transport and Road Research Laboratory, Imperial College of Science and Technology.

Pye, K. and Frinsley, D. H. (1985). Formation of secondary porosity in sandstones by quartz framework grain dissolution. *Nature*, **317**, 54–55.

Raunet, M. (1985). Les bas-fonds en Afrique et a Madagascar. *Zeitschrift für Geomorphologie Supplement Band*, **52**, 25–62.

Read, S. A. L., Richards, L. and Perrin, N. D. (2000). Assessment of New Zealand greywacke rock masses with the Hoek–Brown failure criterion. *Proceedings of GeoEng 2000 International Conference on Geotechnical and Geological Engineering*, 19–24 November 2000, Melbourne.

Reardon, E. J. (1979). Complexing of silica by iron(III) in natural waters. *Chemical Geology*, **25**, 339–345.

Reed, J. C., Bryant, B. and Hack, J. T. (1963). Origin of some intermittent ponds on quartzite ridges in western North Carolina. *Geological Society of America Bulletin*, **74**(9), 1183–1187.

Reiche, P. (1937). The Toreva-block, a distinctive landslide type. *Journal of Geology*, **45**, 538–548.

Richards, B. G., Coulthard, M. A. and Toh, C. T. (1981). Analysis of slope stability at Goonyella mine. *Canadian Geotechnical Journal*, **18**, 179–194.

Rimstidt, J. D. and Barnes, H. L. (1980). The kinetics of silica–water reactions. *Geochimica et Cosmochimica Acta*, **44**, 1683–1699.

Robinson, D. A. and Williams, R. B. G. (1989). Polygonal cracking of sandstone at Fontainebleau, France. *Zeitschrift für Geomorphologie*, **33**, 59–72.

Robinson, D. A. and Williams, R. B. G. (1992). Sandstone weathering in the High Atlas, Morocco. *Zeitschrift für Geomorphologie*, **36**, 413–429.

Robinson, D. A. and Williams, R. B. G. (1994). Sandstone weathering and landforms in Britain and Europe. In *Rock Weathering and Landform Evolution*, eds D. A. Robinson and R. B. G. Williams. Chichester: John Wiley & Sons, pp. 371–391.

Robinson, D. A. and Williams, R. B. G. (2000). Accelerated weathering of a sandstone in the High Atlas Mountains of Morocco by an epilithic lichen. *Zeitschrift für Geomorphologie*, **44**, 513–528.

Robinson, D. A. and Williams, R. B. G. (2005). Comparative morphology and weathering characteristics of sandstone outcrops in England, UK. *Sandstone Landscapes in Europe – Past, Present and Future. Proceedings of the 2nd Conference on Sandstone Landscapes, 25–28.05.2005*, Vianden (Luxembourg), Ferrantia 44, 249 p.

Robinson, E. R. (1970). Mechanical disintegration of the Navajo sandstone in Zion Canyon, Utah. *Geological Society of America Bulletin*, **81**, 2799–2806.

Roland, N. W. (1973). Die Anwendung der Photointerpretation zur Loesung stratigraphischer und tektonischer Probleme im Bereich von Bardai und Aozou (Tibesti-Gebirge, Zentral-Sahara). *Berliner Geographische Abhandlungen*, **19**, 496–506.

Rudnicki, J. W. (1980). Fracture mechanics applied to the Earth's crust. *Annual Review of Earth and Planetary Sciences*, **8**, 489–525.

Rust, I. C. (1981). Lower Paleozoic rocks of South Africa. In *Lower Paleozoic of the Middle East, East and Southern Africa and Antarctica*, ed. C. H. Hallack. Chichester: Wiley, pp. 165–187.

Rutford, R. H. (1972). Glacial geomorphology of the Ellsworth Mountains. In *Antarctic Geology and Geophysics*, ed. R. J. Adie. Oslo: Universitetsforlaget, pp. 225–232.

Sauer, C. O. (1941). Foreword to historical geography. *Annals of the Association of American Geographers*, **31**, 1–24.

Schipull, v. K. (1978). Waterpockets (Opferkessel) in Sandsteinen des Zentralen Colorado-Plateaus. *Zeitschrift für Geomorphologie*, **22**, 426–438.

Schmidt, K. H. (1994). The groundplan of cuesta scarps in dry regions as controlled by lithology and structure. In *Rock Weathering and Landform Evolution*, eds D. A. Robinson and R. B. G. Williams. Chichester: John Wiley & Sons, pp. 355–368.

Schmitthenner, H. (1925). Die Enstehung der Dellen und ihre morphologische Bedeutung. *Zeitschrift für Geomorphologie*, **7**, 3–28.

Schumm, S. A. and Chorley, R. J. (1964). The fall of Threatening Rock. *American Journal of Science*, **262**, 1041–1054.

Schumm, S. A. and Chorley, R. J. (1966). Talus weathering and scarp recession in the Colorado plateaus. *Zeitschrift für Geomorphologie*, **10**, 11–36.

Selby, M. J. (1980). A rock mass strength classification for geomorphic purposes: with tests from Antarctica and New Zealand. *Zeitschrift für Geomorphologie*, **24**, 31–51.

Selby, M. J. (1982). Controls on the stability and inclinations of hillslopes formed on hard rock. *Earth Surface Processes and Landforms*, **7**, 449–467.

Selby, M. J. and Hodder, A. P. W. (1993). *Hillslope Materials and Processes*. Oxford University Press: Oxford.

Self, C. A. and Mullan, G. J. (2005). Rapid karst development in an English quartzitic sandstone. *Acta Carstologica*, **34**(2), 415–424.

Serezhinikov, A. I. (1989). Silica in acid natural solutions. *Transactions (Doklady) of the USSR Academy of Sciences: Earth Science Section*, **298**, 134–138.

Shade, B., Alexander, S. C., Alexander, E. C. J. and Truong, H. (2000). Solutional processes in silicate terranes: True karst vs pseudokarst with emphasis on Pine County, Minnesota [in abstract]. *Abstracts with Programs: 2000 GSA Annual Meeting*, **32**(7), A–27.

Shakesby, R. A., Wallbrink, P. J., Doerr, S. H., Chafer, C. J., Humphreys, G. S., Blake, W. H. and Tomkins, K. M. (2007). Distinctiveness of wild fire effects on soil erosion in south-east Australian eucalypt forests assessed in a global context. *Forest Ecology and Management*, **238**, 347–364.

Shanmugam, G. and Higgins, J. B. (1988). Porosity enhancement from chert dissolution beneath Neocomian unconformity: Ivishak Formation, North Slope, Alaska. *The American Association of Petroleum Geologists Bulletin*, **72**, 523–535.

Siever, R. (1962). Silica solubility, 0–200 °C, and the diagenesis of siliceous sediments. *Journal of Geology*, **70**, 127–150.

Simonett, D. S. (1967). Landslide distribution and earthquakes in the Bewani and Torricelli mountains, New Guinea, statistical analysis. In *Landform Studies from Australia and New Guinea*, eds J. N. Jennings and J. A. Mabbut. Canberra: ANU Press, pp. 64–84.

Skinner, B. J. (1966). Thermal expansion. In *Handbook of Physical Constants*, ed. S. P. Clark. Washington: Geological Society of America, pp. 75–96.

Šmída, B., Audy, M. and Mayoral, F. (2005). Cueva Charles Brewer: Largest quartzite cave in the world. *NSS News*, **63**, 13–31.

Smith, B. J. (1977). Rock temperature measurements from the northwest Sahara and their implications for rock weathering. *Catena*, **4**, 41–63.

Smith, B. J., McGreevy, J. P. and Whalley, W. B. (1987). Silt production by weathering of a sandstone under hot arid conditions: an experimental study. *Journal of Arid Environments*, **12**, 199–214.

Southwick, D. L., Morey, G. B. and Mossler, J. H. (1986). Fluvial origin of the early Proterozoic Sioux Quartzite, southwestern Minnesota. *Geological Society of America Bulletin*, **97**, 1432–1444.

Sparks, B. W. (1971). *Rocks and Relief*. London: Longmans.

Spate, A. (1999). Sandstone landforms. In *Ngarrabullgan: Geographical Investigations in Djungan Country, Cape York Peninsula*, ed. B. David. Monash Publications in Geography and Environmental Science, 51. Clayton: Monash University, pp. 72–77.

Spicker, E. M. (1949). Sedimentary facies and associated diastrophism in the Upper Cretaceous of central and eastern Utah. *Geological Society of America Memoir*, **39**, 55–81.

Sponholz, B. (1989). Karsterscheinungen in nichtkarbonatischen Gesteinen in der östlichen Republik Niger. *Würzburger Geographische Arbeiten*, **75**, 1–265.

Sponholz, B. (2003). Sandstone karst and palaeo-sandstone weathering; its palaeoenvironmental implication and Holocene impact on groundwater flow (Abstract). *XVI INQUA Congress; Programs with Abstracts – Congress of the International Union for Quaternary Research*, Würzburg.

Stacey, T. R. (1970). The stresses surrounding open-pit mine slopes. In *Planning Open Pit Mines*, ed. P. W. van Rensburg. Cape Town: Balkema.

Stacey, T. R. (1974). The behaviour of two- and three-dimensional model rock slopes. *Quaterly Journal of Engineering Geology*, **8**, 67–72.

Stallard, R. E. (1985). River chemistry, geology, geomorphology and soils in the Amazon and Orinoco basins. In *The Chemistry of Weathering*, ed. J. Drever. Dordrecht: Reidel, pp. 293–316.

Steel, R. J. (1976). Devonian basins of western Norway – sedimentary response to tectonism and varying tectonic context. In *Sedimentary Basins of Continental Margins and Cratons*, ed. M. H. Bott. Amsterdam: Elsevier, pp. 207–224.

Stephansson, O. (1971). Stability of single openings in horizontally bedded rock. *Engineering Geology*, **5**, 5–77.

Stern, T. A., Baxter, A. K. and Barrett, P. J. (2005). Isostatic rebound due to glacial erosion within the Transantarctic Mountains. *Geology*, **33**, 221–224.

Strand, T. and Kulling, O. (1972). *Scandinavian Caledonides*. London: Wiley.

Stretch, R. C. and Viles, H. A. (2002). The nature and rate of weathering by lichens on lava flows in Lanzarote. *Geomorphology*, **47**, 87–94.

Sugden, D. E. (1968). The selectivity of glacial erosion in the Cairngorm Mountains. *Transactions of the Institute of British Geographers*, **45**, 79–92.

Sugden, D. and Denton, G. (2004). Cenozoic landscape evolution of the Convoy Range to MacKay Glacier area, Transantarctic Mountains: onshore to offshore synthesis. *Geological Society of America Bulletin*, **116**, 840–857.

Suggate, R. P. (1982). The geological perspective. In *Landforms of New Zealand*, eds J. M. Soons and M. J. Selby. Auckland: Longman Paul, pp. 1–13.

Sullivan, M. E. and Hughes, P. J. (1983). The geoarchaeology of the Sydney Basin sandstones. In *Aspects of Australian Sandstone Landscapes*, eds R. W. Young and G. C. Nanson. Wollongong: Australian and New Zealand Geomorphology Group, pp. 120–126.

Summerfield, M. A., Stuart, F. M., Cockburn, H. A. P., Sugden, D. E., Denton, G. H., Dunait, T., Marchant, D. R. and Harbor, J. (1999). Long-term rates of denudation in the Dry Valleys, Transantarctic Mountains, southern Victoria Land, Antarctica, based on in-situ-produced cosmogenic ^{21}Ne. *Geomorphology*, **27**, 113–129.

Suppe, J. (1987). The active Taiwan mountain belt. In *The Anatomy of Mountain Ranges*, eds J. P. Schacr and J. Rodgers. Princeton: Princeton University Press, pp. 277–293.

Swierkosz, K. (2002). The Ishalo – the gem of Madagascar. *Abstracts of the Conference on Sandstone Landscapes: Diversity, Ecology and Conservation, 14–29 September 2002*, Doubice, Czech Republic.

Szczerban, E., Urbani, F. and Colvée, P. (1977). Cuevas y simas en cuarcitas y metalimolitas del grupo Roraima, Meseta de Guaiquinima Estado Bolivar. *Boletin Sociedad Venezuela Espeleologia*, **8**(16), 127–154.

Szevtes, G. (1989). *Sandstone caves in Nigeria*. Frankfurt am Main.

Tate, G. H. H. (1938). Notes on the Phelps Venezuelan Expedition. *Geographical Review*, **28**, 452–474.

Taylor, G. A., Eggleton, C. C., Holzauer, L. A., Maconachie, M., Gordon, M. C., Brown, M. C. and McQueen, R. G. (1992). Cool climate lateritic and bauxite weathering. *Journal of Geology*, **100**, 669–677.

Termier, H. and Termier, G. (1963). *Erosion and Sedimentation*. London: Van Nostrand.

Teruta, Y. (1963). Recherche sur l'agile de Prek Thnot, Cambodge. *Applied Geography*, **4**, 27–37.

Terzaghi, K. (1962). Stability of steep slopes on hard unweathered rock. *Geotechnique*, **12**, 251–270.

The Natural Arch and Bridge Society. (2008). 'The Natural Arch and Bridge Society'. Retrieved 21/12/07, from http://www.naturalarches.org.

Thiry, M. (2005). Morphologies of the Fontainebleau Sandstones and related silica mobility. *Sandstone Landscapes in Europe – Past, Present and Future. Proceedings of the 2nd Conference on Sandstone Landscapes, 25–28.05.2005*, Vianden (Luxembourg), Ferrantia 44, 249 p.

Thiry, M. (2006). Fontainebleau Sandstone: from silicification to pseudokarst development and their use as shelters and carving walls. *Abstracts of the 9th International Symposium on Pseudokarst, 24–26 May 2006*, Bartkowa, Poland, Institute of Nature Conservation, Polish Academy of Sciences, Cracow.

Thiry, M., Ayrault, M. B. and Grisoni, J. C. (1988). Groundwater silicification & leaching in sands: example of the Fontainebleau Sand in the Paris Basin. *Geological Society of America Bulletin*, **100**, 1283–1290.

Thiry, M. and Liron, M. N. (2008). Fontainebleau sandstones (France). In *Sandstone Landscapes*, eds H. Hartel, V. Cilek, T. Herben, A. Jackson and R. Williams. New York: Academia, pp. 359–361.

Thomas, H. H. (1976). *The Engineering of Large Dams*. London: Wiley.

Thomas, M. F. (1974). *Tropical Geomorphology*. London: Macmillan.

Thomas, M. R. and Goudie, A. S. (1985). Dambos: small channelless valleys in the tropics – preface. *Zeitschrift für Geomorphologie Supplement Band*, **52**, v–vi.

Thornbury, W. D. (1965). *Regional Geomorphology of the United States*. New York: Wiley.

Tinkler, K. J. (1993). Fluvially sculpted rock bedforms in Twenty Mile Creek, Niagara Peninsula, Ontario. *Canadian Journal of Earth Science*, **30**, 945–953.

Tippett, J. M. and Kamp, P. J. J. (1995). Geomorphic evolution of the Southern Alps, New Zealand. *Earth Surface Processes and Landforms*, **20**, 177–192.

Tricart, J. (1974). *Structural Geomorphology*. London: Longman.

Tricart, J. (1979). Alteration et dissection differentielles dans un socle cratonise. *Annales de Geograpie*, **487**, 265–313.

Tricart, J. and Cailleux, A. (1972). *Introduction to Climatic Geomorphology*. London: Longmans.

Tripathi, J. K. and Rajamani, V. (2003). Weathering control over geomorphology of supermature Proterozoic Delhi quartzites of India. *Earth Surface Processes and Landforms*, **28**, 1379–1387.

Turkington, A. V. and Paradise, T. R. (2005). Sandstone weathering: a century of research and innovation. *Geomorphology*, **67**, 229–253.

Twidale, C. R. (1956). Der 'Bienenkorb' eine neue morphologische form aus Nord-Queensland, Nord-Australien. *Erdkunde*, **10**, 239–240.

Twidale, C. R. (1967). Hillslopes and pediments in the Flinders Ranges, South Australia. In *Landform Studies from Australia and New Guinea*, eds J. N. Jennings and J. A. Mabbutt. Canberra: ANU Press, pp. 95–117.

Twidale, C. R. (1971). *Structural Landforms*. Canberra: ANU Press.

Twidale, C. R. (1973). On the origin of sheet jointing. *Rock Mechanics*, **5**, 163–187.

Twidale, C. R. (1978). On the origin of Ayers Rock, central Australia. *Zeitschrift für Geomorphologie Supplement Band*, **31**, 177–206.

Twidale, C. R. (1980). Origin of minor sandstone landforms. *Erdkunde*, **34**, 219–224.

Twidale, C. R. (1982). *Granite Landforms*. Amsterdam: Elsevier.

Twidale, C. R. (1983). Pediments, peneplains and ultiplains. *Revue de Geomorphologie Dynamique*, **32**, 1–35.

Twidale, C. R. (1987). Sinkholes (Dolines) in lateritised sediments, Western Sturt Plateau, Northern Territory, Australia. *Geomorphology*, **1**, 33–52.

Twidale, C. R. and Campbell, E. M. (2005). *Australian Landforms*. Dutal: Rosenberg.

Twidale, C. R. and Harris, W. K. (1977). On the age of Ayers Rock and the Olgas, central Australia. *Transactions of the Royal Society of South Australia*, **101**, 45–50.

Uba, C. E., Heubeck, C. and Hulka, C. (2005). Facies analyses and basin architecture of the Neogene Subandean synorogenic wedge, southern Bolivia. *Sedimentary Geology*, **180**, 91–123.

Underhill, J. R. and Woodcock, N. H. (1987). Faulting mechanisms in high-porosity sandstones, New Red Sandstone, Arran, Scotland. In *Deformation of Sediments and Sedimentary Rocks*, eds M. E. Jones and R. M. Preston. Oxford: Blackwell, pp. 91–105.

Urban, J., Margielewski, W., Alexandrowicz, Z. and Mleczek, T. (2006). Excursion Guide Book. *Abstracts of the 9th International Symposium on Pseudokarst*, 24–26

May 2006, Bartkowa, Poland, Institute of Nature Conservation, Polish Academy of Sciences, Cracow.

Urban, J., Margielewski, W., Zak, K., Schejbal-Chwastek, M., Mleczek, T., Szura, C., Hercman, H. and Sujka, G. (2006). Preliminary data on speleothems in the caves of the Beskidy Mts., Poland. *Abstracts of the 9th International Symposium on Pseudokarst*, 24–26 May 2006, Bartkowa, Poland, Institute of Nature Conservation, Polish Academy of Sciences, Cracow.

Urbani, F. (1976). Opalo, calcedonia y calcita en la Cueva del Cerro Autano (Am 11), Territorio Federal Amazonas, Venezuela (in Spanish, English summary). *Boletin Sociedad Venezuela Espeleologia*, **7**(14), 129–145.

Urbani, F. (1977). Novedades sobre estudios realizados en las formas carsicas y pseudocarsicas del Escud de Guayana (in Spanish, English summary). *Boletin Sociedad Venezuela de Espeleologia*, **8**(16), 175–197.

Urbani, F. (1990). Algunos comentarios sobre terminologia karstica aplicada a rocas siliceas. *Boletin Sociedad Venezuela Espeleologia*, **24**, 5–6.

Urbani, F. (1996). Venezuelan cave minerals: a review. *Boletín de la Sociedad Venezolana de Espeleología*, **30**, 1–13.

Urbani, F. and Szczerban, E. (1974). Venezuelan caves in non-carbonate rocks: a new field in karst research. *NSS News*, **32**, 233–235.

Urbanova, H. and Prochazka, J. (2005). Kokorínsko protected landscape area – Rare species, protection and conservation. *Sandstone Landscapes in Europe – Past, Present and Future. Proceedings of the 2nd Conference on Sandstone Landscapes, 25–28.05.2005*, Vianden (Luxembourg), Ferrantia 44, 249 p.

US Geological Survey. (undated). 3-D Reservoir Characterization of the House Creek Oil Field, Powder River Basin, Wyoming, V1.00. *U.S. Geological Survey Digital Data Series DDS-33.* http://pubs.usgs.gov/dds/dds-033/USGS_3D/ssx.txt/geology.htm (accessed 6 December 2007).

Varilová, Z. (2002). Geomorphology of the Bohemian Switzerland National Park, Czech Republic. *Abstracts of the Conference on Sandstone Landscapes: Diversity, Ecology and Conservation, 14–20 September 2002*, Doubice in Saxonian-Bohemian Switzerland, Czech Republic.

Vdovets, M. (2006). Pseudokarst and geoconservation in Russia. *Abstracts of the 9th International Symposium on Pseudokarst*, 24–26 May 2006, Bartkowa, Poland, Institute of Nature Conservation, Polish Academy of Sciences, Cracow.

Velbel, M. A. (1985). Hydrochemical constraints on mass balances in forested watersheds of the Southern Appalachians. In *The Chemistry of Weathering*, ed. J. Drever. Dordrecht: Reidel, pp. 231–247.

Verges, J., Marzo, M. and Munoz, J. A. (2002). Growth strata in foreland settings. *Sedimentary Geology*, **146**, 1–9.

Vidal Romani, J. R. and Twidale, C. R. (1999). Sheet fractures, other stress forms and some engineering implications. *Geomorphology*, **31**, 13–27.

Viles, H. A. and Goudie, A. S. (2004). Biofilms and case hardening on sandstones from Al-Quwayra, Jordan. *Earth Surface Processes and Landforms*, **29**, 1473–1485.

Viles, H. A. and Pentecost, A. (1994). Problems in assessing the weathering action of lichens with an example of epiliths on sandstone. In *Rock Weathering and Landform Evolution*, eds D. A. Robinson and R. B. G. Williams. Chichester: John Wiley & Sons.

Vincent, P. and Katton, F. (2006). Yardangs on the Cambro-Ordovician Saq Sandstones, Northwest Saudi Arabia. *Zeitschrift für Geomorphologie*, **50**, 305–320.

Voight, B. (1978). Lower Gros Ventre slide. In *Rock Slides and Avalanches 1. Natural Phenomena*, ed. B. Voight. Amsterdam: Elsevier, pp. 113–166.

von Damm, K. L., Bischoff, J. L. and Rosenbaure, R. J. (1991). Quartz solubility in hydrothermal seawater: an experimental study and equation describing quartz solubility for up to 0.5 M NaCl solutions. *American Journal of Science*, **291**, 977–1007.

von Engeln, O. D. (1940). A particular case of knickpunkte. *Annals of the Association of American Geographers*, **30**, 268–271.

von Engeln, O. D. (1957). *Geomorphology, Systematic and Regional*. New York: MacMillan.

Warscheid, T. and Braams, J. (2000). Biodeterioration of stone: a review. *International Biodeterioration and Biodegradation*, **46**, 343–368.

Watchman, A. (1992). Composition, formation and age of some Australian silica skins. *Australian Aboriginal Studies*, **1**, 61–66.

Watchman, A. (2007). Evidence of a 25,000-year-old pictograph in Northern Australia. *Geochaeology*, **8**, 465–473.

Watson, R. A. and Wright, H. E. (1963). Landslides on the east flank of the Chuska Mountains, northwestern New Mexico. *American Journal of Science*, **261**, 525–548.

Webb, J. A. and Finlayson, B. L. (1984). Allophane and opal speleothems from granite caves in south-east Queensland. *Australian Journal of Earth Science*, **31**, 341–349.

Wellman, P. and McDougall, I. (1974). Potassium–argon ages on the Cainozoic volcanic rocks of New South Wales. *Journal of the Geological Society of Australia*, **21**, 247–272.

Wernick, E., Pastone, E. L. and Pires Neto, A. (1977). Cuevas en Areniscas, Rio Claro, Brasil. *Boletin Sociedad Venezuela Espeleologia*, **8**(16), 1699–1707.

Wessels, D. C. J. and Schoeman, P. (1988). Mechanism and rate of weathering of Clarens sandstone by an endolithic lichen. *South African Journal of Science*, **84**, 275–277.

Whalley, W. B., Douglas, G. R. and McGreevy, J. P. (1982). Crack propagation and associated weathering in igneous rocks. *Zeitschrift für Geomorphologie*, **26**, 33–54.

Whitaker, C. R. (1978). Pediment form and evolution on granite in the East Kimberley. In *Landform Evolution in Australasia*, eds J. L. Davies and M. A. Williams. Canberra: Australian National University Press, pp. 306–330.

White, W. B., Jefferson, G. L. and Haman, J. F. (1966). Quartzite karst in southeastern Venezuela. *International Journal of Speleology*, **2**, 309–316.

Whitehouse, I. E. (1983). Distribution of large rock avalanche deposits in the central Southern Alps, New Zealand. *New Zealand Journal of Geology and Geophysics*, **26**, 271–279.

Wiegand, J., Fey, M., Haus, N. and Karmann, I. (2004). Geochemische und hydrochemische Untersuchung zur Genese von Sandstein- und Quarzitkarst in der Chapada Diamantina und in Eisernen Viereck (Brasilien). *Zeitschrift der Deutschen Geologischen Gesellschaft*, **155**, 61–90.

Wilkinson, M. T., Chappell, J., Humphreys, G. S., Fifield, K., Smith, B. and Hesse, P. (2005). Soil production in heath and forest, Blue Mountains, Australia: influence of lithology and palaeoclimate. *Earth Surface Processes and Landforms*, **30**, 923–934.

Willemin, J. H. (1984). Erosion and the mechanics of shallow foreland thrusts. *Journal of Structural Geology*, **6**, 425–432.

Willems, L., Compère, P., Hatert, F., Pouclet, A., Vicat, J. P., Ek, C. and Boulvain, F. (2002). Karst in granitic rocks, South Cameroon: cave genesis and silica and taranakite speleothems. *Terra Nova*, **14**, 355–362.

Willems, L., Compere, P. and Sponholz, B. (1998). Study of siliceous karst genesis in eastern Niger: microscopy and X-ray microanalysis of speleothems. *Zeitschrift für Geomorphologie*, **42**, 129–142.

Willems, L., Poudet, A., Lenoir, F. and Vicat, J. P. (1996). Phénomènes karstiques en milieux non-carbonatés. Etude de cavités et problématique de leur développement au Niger Occidental. *Zeitschrift für Geomorphologie Supplement Band*, **103**, 193–214.

Williams, M. A. J. (1973). The efficacy of creep and slopewash in tropical and temperate Australia. *Australian Geographical Studies*, **11**, 62–78.

Williams, R. B. G. and Robinson, D. A. (1989). Origin and distribution of polygonal cracking of rock surfaces. *Geografisker Annaler*, **71-A**, 145–159.

Wilson, P. (1979). Experimental investigation of etch pit formation on quartz sand grains. *Geological Magazine*, **116**, 477–482.

Winkler, E. M. (1975). *Stone Properties, Durability in Man's Environment*. Wien: Springer.

Winkler, E. M. (1994). *Stone in Architecture: Properties, Durability*. Berlin: Springer.

Withe, A. F. and Peterson, M. (1990). The role of reactive surface areas in chemical weathering. *Chemical Geology*, **84**, 334–336.

Wohl, E. E. (1993). Bedrock channel incision along Piccaninny Creek, Australia. *Journal of Geology*, **101**, 749–761.

Wohl, E. E. (1998). Bedrock channel morphology in relation to erosional processes. In *Rivers over Rock, Fluvial Processes in Bedrock Channels*, eds K. J. Tinkler and E. E. Wohl. Washington: American Geophysical Union Monograph Series, pp. 133–151.

Wray, R. A. L. (1996). The morphology and genesis of drainage runnels on the Sydney Basin quartz sandstones. *Proceedings of the 7th Meeting of the Australian and New Zealand Geomorphology Group*, Cairns, Qld.

Wray, R. A. L. (1997a). Quartzite dissolution: karst or pseudokarst? *Cave and Karst Science, Transactions of the British Cave Research Association*, **24**(2), 81–86.

Wray, R. A. L. (1997b). A global review of solutional weathering forms on quartz sandstone. *Earth Science Reviews*, **42**, 137–160.

Wray, R. A. L. (1997c). The formation and significance of coralline silica speleothems in the Sydney Basin, southeastern Australia. *Physical Geography*, **18**, 1–16.

Wray, R. A. L. (1999). Opal and chalcedony speleothems on quartz sandstones in the Sydney region, southeastern Australia. *Australian Journal of Earth Sciences*, **46**, 623–632.

Wright, H. E. (1964). Origin of the lakes in the Chuska Mountains, Northwestern New Mexico. *Geological Society of America Bulletin*, **75**, 589–598.

Wylie, D. C. (1980). Toppling rock slope failures, examples of analysis and stabilization. *Rock Mechanics*, **13**, 89–98.

Yanes, C. E. and Briceño, H. O. (1993). Chemical weathering and the formation of pseudo-karst topography in the Roraima Group, Gran Sabana, Venezuela. *Chemical Geology*, **107**, 341–343.

Yariv, S. and Cross, H. (1979). *Geochemistry of Colloid Systems*. Berlin: Springer.

Yatsu, E. (1966). *Rock Control in Geomorphology*. Tokyo: Sozosha.

Yatsu, E. (1988). *The Nature of Weathering*. Tokyo: Sozosha.

Young, A. (1972). *Slopes*. Edinburgh: Oliver and Boyd.

Young, A. R. M. (1977). The characteristics and origin of coarse debris deposits near Wollongong, Australia. *Catena*, **4**, 289–307.

Young, A. R. M. (1986). The geomorphic development of dells (upland swamps) on the Woronora Plateau, N.S.W., Australia. *Zeitschrift für Geomorphologie*, **30**, 317–327.

Young, A. R. M. (1987). Salt as an agent in the development of cavernous weathering. *Geology*, **15**, 962–966.

Young, A. R. M. and Young, R. W. (2001). *Soils in the Australian Landscape*. Melbourne: Oxford.

Young, R. G. (1955). Sedimentary facies and intertonguing in the Upper Cretaceous of the Book Cliffs, Utah–Colorado. *Geological Society of America Bulletin*, **66**, 171–201.

Young, R. W. (1977a). Landscape development in the Shoalhaven River catchment of southeastern New South Wales. *Zeitschrift für Geomorphologie*, **21**, 262–283.

Young, R. W. (1977b). Geological and hydrological influences on the development of meandering valleys in the Shoalhaven River catchment, southeastern Australia. *Erdkunde*, **32**, 171–182.

Young, R. W. (1983a). Sandstone terrain in a semi-arid littoral environment: the lower Murchison valley, Western Australia. *Australian Geographer*, **17**, 143–153.

Young, R. W. (1983b). Block gliding in sandstones of the southern Sydney Basin. In *Aspects of Australian Sandstone Landscapes*, eds R. W. Young and G. C. Nanson. Wollongong: Australian and New Zealand Geomorphology Group, pp. 31–38.

Young, R. W. (1983c). The tempo of geomorphological change; evidence from southeastern Australia. *Journal of Geology*, **91**, 221–230.

Young, R. W. (1985). Waterfalls: form and process. *Zeitschrift für Geomorphologie Supplement Band*, **55**, 81–95.

Young, R. W. (1986). Tower karst in sandstone: Bungle Bungle massif, northwestern Australia. *Zeitschrift für Geomorphologie*, **30**, 189–202.

Young, R. W. (1987). Sandstone landforms of the tropical East Kimberley region, northwestern Australia. *Journal of Geology*, **95**, 205–218.

Young, R. W. (1988). Quartz etching and sandstone karst; examples from the east Kimberleys, northwestern Australia. *Zeitschrift für Geomorphologie*, **32**, 409–423.

Young, R. W. and McDougall, I. (1985). The age, extent and geomorphological significance of the Sassafras basalt, south-eastern New South Wales. *Australian Journal of Earth Sciences*, **32**, 323–331.

Young, R. W. and White, K. L. (1994). Satellite imagery analysis of landforms: illustrations from southeastern Australia. *Geocarto International*, **9**(2), 33–44.

Young, R. W. and Wray, R. A. L. (2000). Contribution to the theory of scarpland development from observations in central Queensland, Australia. *Journal of Geology*, **108**, 705–771.

Young, R. W., Wray, R. A. L., White, K. L. and Hoskins, E. (1995). Rheological deformation and block gliding in the southern Sydney Basin. *Australian Geographer*, **26**, 180–188.

Young, R. W. and Young, A. R. M. (1988). "Altogether barren, peculiarly romantic": The sandstone lands around Sydney. *The Australian Geographer*, **19**, 9–25.

Young, R. W. and Young, A. R. M. (1992). *Sandstone Landforms*. Berlin: Springer.

Youngson, J., Bennett, E., Jackson, J., Norris, R., Raisbeck, G. and Yiou, F. (2005). 'Sarsen Stones' at German Hill, Central Otago, New Zealand, and their potential for in situ cosmogenic isotope dating of landscape evolution. *Journal of Geology*, **113**, 341.

Youngson, J. H. (2005). Diagenetic silcrete and formation of silcrete ventifacts and aeolian gold placers in Central Otago, New Zealand. *New Zealand Journal of Geology and Geophysics*, **48**, 247–263.

Zawidzki, P., Urbani, F. and Koisar, B. (1976). Preliminary notes on the geology of the Sarisari-ama Plateau, and the origin of its caves. *Boletín de la Sociedad Venezolana de Espeleología*, **7**(13), 29–37.

Zuo, D. and Xing, Y. (1992). *The Natural Features of China*. Beijing: China Pictorial Publishing.

Index

Printed in the United States
By Bookmasters